REALM OF ICE
AND SKY

ALSO BY BUDDY LEVY

Empire of Ice and Stone: The Disastrous and Heroic Voyage of the Karluk

Labyrinth of Ice: The Triumphant and Tragic Greely Polar Expedition

No Barriers: A Blind Man's Journey to Kayak the Grand Canyon

Geronimo: Leadership Strategies of an American Warrior

River of Darkness: Francisco Orellana and the Deadly First Voyage Through the Amazon

Conquistador: Hernán Cortés, King Montezuma, and the Last Stand of the Aztecs

American Legend: The Real-Life Adventures of David Crockett

Echoes on Rimrock: In Pursuit of the Chukar Partridge

REALM OF ICE AND SKY

TRIUMPH, TRAGEDY, AND HISTORY'S GREATEST ARCTIC RESCUE

BUDDY LEVY

ST. MARTIN'S PRESS
NEW YORK

First published in the United States by St. Martin's Press, an imprint of St. Martin's Publishing Group

www.stmartins.com

Library of Congress Cataloging-in-Publication Data

Names: Levy, Buddy, 1960– author.
Title: Realm of ice and sky : triumph, tragedy, and history's greatest Arctic rescue / Buddy Levy.
Description: First edition. | New York : St. Martin's Press, 2025. | Includes bibliographical references and index.
Identifiers: LCCN 2024031110 | ISBN 9781250289186 (hardcover) | ISBN 9781250289193 (ebook)
Subjects: LCSH: Arctic regions—Discover and exploration. | Arctic regions—Discovery and exploration—Italian. | Nobile, Umberto, 1885–1978. | Arctic regions—Aerial exploration—History. | Airships—Accidents—Arctic regions. | Italia (Airship) | Search and rescue operations—Arctic regions—History—20th century.
Classification: LCC G700 1928 .L48 2025 | DDC 910.911/3—dc23/eng/20240910
LC record available at https://lccn.loc.gov/2024031110

Our books may be purchased in bulk for promotional, educational, or business use. Please contact your local bookseller or the Macmillan Corporate and Premium Sales Department at 1-800-221-7945, extension 5442, or by email at MacmillanSpecialMarkets@macmillan.com.

First Edition: 2025

10 9 8 7 6 5 4 3 2 1

*For my granddaughter Palmer Drew—
the newest adventurer in the family*

CONTENTS

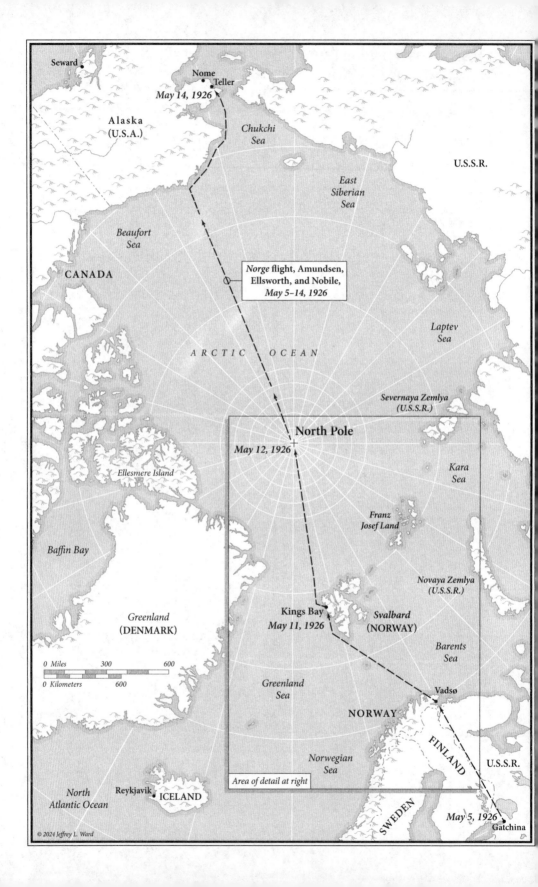

AIRSHIP ROUTES OF WELLMAN, AMUNDSEN, ELLSWORTH, NOBILE, AND THE RESCUE ROUTES OF THE ICEBREAKER *KRASSIN*

Severnaya Zemlya
(U.S.S.R.)

May 24, 1928 **North Pole**

Italia flight, Nobile,
May 15–18, 1928

*Kara
Sea*

A R C T I C   O C E A N

Italia flight, Nobile,
May 23–25, 1928

*Franz
Josef Land*

Italia crash site,
May 25, 1928 ✕

Foyn

*Greenland
(DENMARK)*

Broch

Charles XII Is.

*Novaya Zemlya
(U.S.S.R.)*

Danes Island, origin
of Wellman's flight
(see detail map below)

*Svalbard
(NORWAY)*

Kings Bay

0 Miles   100   200   300

0 Kilometers   300

KINGS BAY—Origin of the
Flights of the *Norge* (1926)
and *Italia* (1928)

*Barents
Sea*

*Greenland
Sea*

Italia route, Nobile, from
Rome, Italy to Kings Bay,
arrived *May 5, 1928*

Bear Island

ESTIMATED ROUTE OF
WELLMAN'S POLAR FLIGHT,
SEPT. 2, 1907

Amundsen's final flight,
June 18, 1928

Vadsø

ARCTIC OCEAN

U.S.S.R.

0 Mi.   5

0 Km.   5

*Smeerenburg
Point*

Fowl Point

Krassin Expedition,
June 24–July 19, 1928

Tromsø

*Amsterdam
Island*

Fowl Bay

Virgo Bay

Svitjod Glacier

*Norwegian
Sea*

FINLAND

*Danes
Island*

Svalbard

NORWAY

SWEDEN

© 2024 Jeffrey L. Ward

PRELUDE

TO RISE FROM THE EARTH INTO THE SKY AND SOAR FREELY ALOFT had been dreamed of and desired since humans first began watching birds wing and hover overhead, their effortless dance in the wind a wonder and an envy. Over the centuries, from the Greek mythology of Daedalus and Icarus to Leonardo da Vinci, the ideas and visions of how humans might harness the air took myriad forms, depicted in stories and drawings and even in fabricated prototypes.

At the dawn of the twentieth century, manpowered flight was in its fledgling stage. Balloons of various types, including hot air and hydrogen, had been flown in France for the last hundred years, but these could not really be steered or navigated, and were thus at the mercies and whims of the wind. Pure, true flight would require power and the ability to control direction.

By the early 1900s, the free-floating balloon was about to be supplanted by two new experiments, the airship (motorized, lighter-than-air balloons, also called dirigibles) and the airplane (at the time, "aeroplane"). Airships developed first, gaining international attention when wealthy Parisian businessman M. Deutsch de La Meurthe (the so-called "Oil King of Europe") began, in 1900, offering a prize of 100,000 francs for the first motor-propelled dirigible that could fly from the Aero Club grounds at St. Cloud around the Eiffel Tower and back—a distance of seven miles—within thirty minutes. Thousands of fascinated French and international spectators and journalists came to watch these competitions, and a global airship craze ensued.

During the same period, in a much less public way, the Wright brothers were testing powered, winged, and piloted "heavier-than-air" flight in their *Wright Flyer* in late 1903. They flew 120 feet in twelve seconds on their first successful flight, 175 feet on their second, and 200 feet on their third. By then, the largest dirigibles being built by Germany's Count Ferdinand von Zeppelin were over four hundred feet long. With the first airplane flying only half the length of the largest dirigible, at this moment in aeronautical history, the airship was clearly orders of magnitude ahead in the flight race.

Into this exciting aerial milieu, in a strange but not entirely unexpected confluence, entered the Arctic explorer and the race for the North Pole. The North Pole remained the holy grail of human discovery, one of the last remaining blank spots on the world's maps. During the period that this book covers (1900–1928, primarily), the North Pole had yet to be reached. Land-based and ship-based attempts, nearly all using teams of sled dogs, had all come up short, always after terrible misery and sometimes at the cost of lost limbs and lives.

In the span of three decades, three men—American Walter Wellman, Norwegian Roald Amundsen, and Italian Umberto Nobile—would try to fill in that blank spot on the map at the top of the world, and they would do so not by the traditional, long-tried and always failed dogsled method, but by airship. *Realm of Ice and Sky* is a serial history of the aerial explorations to reach the summit of the Earth. It is not the history of a single voyage but rather is a history of a single *type* of voyage—airborne, in experimental "lighter than air" flying machines called airships or dirigibles.

The aerial explorers would innovate and experiment and risk their own lives and the lives of their crews in a dangerous, deadly quest to be first to fly to the North Pole.

There would be joy and sorrow, tragedy and suffering.

There would be disaster.

There would be triumph.

PROLOGUE

Aerial navigation will solve the mystery of the North Pole and the frozen ocean.
—Walter Wellman, 1893

Camp Wellman. Danes Island (Danskøya),
northwest Spitsbergen—September 2, 1907

WHITECAPS LASHED AGAINST THE ICY SHORELINE OF NORWAY'S
Smeerenburg Sound. Beyond the roiling bay, knobby peaks thrust from
the sea up to four thousand feet, their craggy faces lined with snow. Aqua-
marine icebergs—some carved into archways and tunnels, some like spired
castles—drifted shoreward, driven by the wind. In the mountain valleys,
great glaciers groaned and rumbled, cascading to the sea, calving into the
water with echoing roars.

The dim light of early morning revealed an enormous oblong form
rising from the bay like an ominous gray cloud, but it was nothing that
came from land or sea. It was a man-made thing—a giant gas-filled balloon emerging from a massive wooden hangar, guided out by three dozen
men clinging to taut draglines and cables. It was the airship *America,* a
spherical, 185-foot-long, 50-foot-wide dirigible inflated with 274 thousand
cubic feet of highly flammable hydrogen, powered by a single sputtering 750-pound, 75-horsepower gasoline engine. In the airship's gondola
rode three men: the visionary expedition leader Walter Wellman; engineer
Melvin Vaniman; and navigator Felix Riesenberg. They were attempting
something perilous and untried in history: to fly a motorized airship to the
North Pole, and to somehow, with luck, return alive.

The men, all Americans, were exhilarated by the prospect of being first to the North Pole, an achievement promising fame, national pride, wealth, and polar immortality. But they were fearful, for the 1,500-mile flight would be *America's* maiden voyage. The airship was untested. Some in the press had called the attempt a "suicide mission," while others lauded the men for their bravery. What everyone knew for certain, however—including the three men aboard the *America*—was that their chances of dying on the journey were likely greater than their chances of surviving it.

Though it was dangerously late in the season to strike out toward the North Pole, the winds were the most favorable they'd been in a month, blowing consistently from the south. Should anything go wrong—and as Walter Wellman knew firsthand from his previous over-ice sledge attempts at the north's holy grail, something always went wrong—he and his crew and dogs would be marooned on moving sea ice with the "Long Night" of polar winter approaching. In the coming months of near-total darkness, they'd be forced to winter over on the grinding pack ice, a terrifying prospect. Few polar explorers had survived winter on the moving, fracturing sea ice hundreds of miles above the Arctic Circle.

Wellman's airship now hovered, fully inflated, its eleven-and-a-half-foot steel propeller whirring. Poised to fly north, Wellman was deeply aware of the risks to himself and his two crewmen. He had explored the Arctic on and off for more than a decade, on various expeditions questing for the North Pole. From his multiple expeditions, he had learned much about the Arctic, including that it would always try, one way or another, to kill you. Nearly fifty years old, Wellman was very lucky to still be alive. On one previous polar expedition, he had fallen into a deep crevasse and had been miraculously rescued. Once, he had been attacked by a polar bear, and was saved only when his sled dogs came snarling to his defense, driving the bear away. Another time, he had broken his shinbone on a jagged ice ridge and faced amputation of his leg, which he refused. He was now reliant on a cane, a grudging reminder of his mortality.

There was nothing typical about Walter Wellman as an Arctic explorer, and his experimental airship was only the most recent example of his unusual nature and innovative ideas and approach. Wellman was an un-likely Arctic explorer in other ways too. He was among the most famous men in the world, a well-connected Washington insider, a celebrated and prolific journalist who frequently hobnobbed with the most powerful men in the country, including President Theodore Roosevelt, who supported his Arctic expeditions. Only recently, Wellman—who wanted to be the first not only to reach the North Pole but also broadcast his achievement to the world in *real time*—had built a series of wireless telegraph stations in the Arctic for that express purpose: one to be installed on the airship *America,* one on Danes Island, and one on mainland Norway.

To illustrate that the new, revolutionary Marconi wireless communication technology would perform this far north, Wellman had sent the following message to President Roosevelt:

ROOSEVELT, WASHINGTON:
GREETINGS, BEST WISHES, BY FIRST WIRELESS MESSAGE
EVER SENT FROM ARCTIC REGIONS.
    WELLMAN

Along with wireless telegraphic communication, Wellman had also experimented with other novel but unproven techniques in polar travel, including using newly invented lightweight aluminum boats for crossing open leads in the sea ice and the use of ultra-concentrated foods for light, high-calorie rations. He had even tested gas-powered "motor-sledges" (dubbed "gasoline dogs") to replace some of the many dogs required for long and arduous sledge journeys.

But it was the motorized airship that was the most experimental, the most pioneering, the most dangerous, and the most controversial of Wellman's many "firsts." No one had yet dared to attempt flying a highly combustible, engine-driven dirigible to the North Pole, for the airship had only recently been invented. Yet Wellman had been planning and scheming since the Brazilian Alberto Santos-Dumont had made history in 1901 (and won that year's 100,000 franc prize!) by flying his 108-foot-long dirigible seven miles over Paris, circling the Eiffel Tower to the cheering of thousands of spectators. The flight was an international sensation, and although Santos-Dumont remained aloft for only thirty minutes, Wellman saw the immense potential of the airship. He had also read every sea captain's logbook and every expedition journal written about the Arctic and attempts to reach the top of the world.

Ten years earlier, the Swedish aeronaut S.A. Andrée and two compatriots had departed from this very spot on Danes Island, on the northwestern coast of Spitsbergen[1] in *Örnen* (*Eagle*), a non-motorized hydrogen-filled balloon. They'd hoped that steady, consistent winds from the south would propel the balloon across the roof of the world and all the way over the pole to Alaska. But on that day in 1897, they had vanished into the misty Arctic air, and no one had seen or heard from them since, except for a few cryptic notes delivered by passenger pigeons early in their flight. Wellman understood he could learn from Andrée's mistakes, first and foremost by adding an engine to his craft, making it a powered, steerable airship rather than a mere passive balloon. He had also consulted with the still-living legends in polar travel, including Norway's hero Fridtjof Nansen and America's Robert Peary, who was at this very moment also preparing for his own ship-based attempt at the North Pole.

It was Robert Peary who was foremost on Wellman's mind as he prepared to cast off into the frozen unknown. Peary, a veteran explorer of the last two decades, had the previous year set the world record for "Farthest North," reaching 87° 6′ N in 1906. Now, the dauntless fifty-one-year-old Peary, who had lost eight toes to frostbite in a previous attempt, was organizing one final voyage, and most in the global exploring community believed Peary had the best chance of winning the prize. Nansen himself had told President Theodore Roosevelt, "Peary is your best man; in fact, I

---

1 Spitsbergen is today generally referred to as Svalbard. Spitsbergen is the largest island within the archipelago of Svalbard. During the time period covered in this book, "Spitsbergen" was often used by chroniclers as the general term for the entire archipelago. Additionally, the Norwegian name for Danes Island is Danskøya; Virgo Harbor is Virgohamna; Amsterdam Island is Amsterdamøya; White Island is Kvitøya; and Bear Island is Bjørnøya.

think he is on the whole the best of the men now trying to reach the Pole, and there is a good chance that he will be the one to succeed."

Wellman begged to differ. Certainly, he respected Robert Peary's accomplishments. They'd met by coincidence at the White House in early December of 1906, each having a separate, private audience with President Roosevelt. The meeting between Wellman and Peary was brief and convivial, but Wellman strongly disagreed with Nansen's assessment of who might reach the North Pole first. He had staked his personal finances, as well as the considerable backing of financiers, his own reputation, the reputation of his newspaper the *Chicago Record-Herald,* and now his life on it. Wellman could see what many others could not: the world was changing. Technology was advancing at a breakneck pace, the last two decades having witnessed the development of wireless communication, the automobile, the airplane, and now the airship. At that moment, it remained anyone's guess as to whether the airplane or the airship would ultimately achieve dominance of the skies for commercial passenger travel. The airships of that time held several advantages, including substantially larger payload capabilities and much greater range. Their potential of a catastrophic fire was not yet fully appreciated, however.

Wellman understood that the race to the pole against Peary stood for much more than just polar bragging rights: it heralded the changing of the guard, and a shift from the traditional man-and-dog over-ice approach to the use of the newest available technologies in polar exploration and scientific discovery. Wellman had said as much: "Recently it seemed . . . the time had come to adopt new methods, to make an effort to substitute modern science for brute force, the motor-driven balloon (airship) for the muscles of men and beasts stumbling along . . . in their heroic struggle to accomplish the almost impossible."

The rivalry with Peary represented something greater still: with instantaneous wireless correspondence, in the age of competing media and big-name newspaper battles, the intrepid reporter stood to make the scoop of the century, one to rival the famous meeting between Stanley and Livingstone in Africa. Should Wellman succeed, he would not only join the pantheon of the greatest polar explorers but would achieve journalistic immortality as well. It was a seductive combination.

For his part, Robert Peary had publicly expressed doubt about Wellman and the use of airships for polar travel, believing such craft too flimsy for the brutal Arctic conditions. Peary had been quoted in *The Washington Post* on May 4, 1907, calling Wellman "a mere interloper in the 'science' of Arctic exploration." Peary also referred to Wellman disparagingly as a "charlatan" and a "hot air voyager."

Wellman had too much pride and bravado to take these insults without response. He fired back to *The New York Times*, "It is not unnatural for one man to have no faith in the scheme of another man, but the difference in this case is that I am acquainted with both the sledging and the airship methods. Commander Peary is probably the most experienced sledger who has started in quest of the pole, but he does not know by actual experience the possibilities of the airship. Personally, I have no faith in the over-the-ice method of reaching the far north as it is now carried on . . . I believe that the airship is the only solution to the problem."

So, the die was cast. With the consent of his two companions Melvin Vaniman and Felix Riesenberg, Walter Wellman was borne aloft in the 115-foot-long steel control car, riding below the great hydrogen-filled envelope, shouting commands over the noise of the wind and the motor. Packed tightly in the control car were also a few sleds, a small boat, food to last ten months, and nearly six thousand pounds of gasoline for the engine. After being walked from the hangar, the *America* was towed into the middle of Smeerenburg Sound by the small steamer *Express* and held steady at an altitude of about five hundred feet above the icy water.

"Cut the line," Wellman yelled, and Felix Riesenberg cast the line off, watching it slither through the air and splash into the water next to the *Express*. *America* soared above the sea, now flying on its own for the first time. "Half-speed," Wellman bellowed, and just above the sound of the engine they heard faint cheers from the crew assembled below at Camp Wellman, who waved and tossed their hats in the air.

"Head her north . . . full speed!" Wellman barked to Riesenberg behind the wheel, and *America* churned through rising mist and fog, heading due north toward the mountains and the polar ice pack beyond. Snow began to fall, the large flakes collecting on the rubberized cotton and silk envelope. From the control car, Wellman looked down from their great height and watched the sea fade away below them, until they were encased in a ghostly expanse of whiteness, disappearing into a frozen realm of ice and sky.

PART ONE

THE PIONEER

The Remarkable
Walter Wellman

1

THE ICE KING WINS

WHEN NOTED AMERICAN JOURNALIST WALTER WELLMAN BOARDED
the airship *America* at Danes Island in 1907 and attempted to fly into
history and polar exploration immortality, he was nearly fifty years old.
Since 1884, he had been obsessed with being the first to reach the North
Pole, and between 1884 and 1907 had traveled to the far north—including
to the archipelagos of Spitsbergen and Franz Josef Land—four times on
that quest. Like many explorers of his day, his efforts had ended with
shipwreck, terrifying ordeals on the moving polar ice, and death to some
expedition members.

His first two forays north—in 1894 and 1898—had been by traditional approach: sail a ship as far north as possible, establish a base (either
on the ship or on land), and then strike out onto the pack ice with dogsled teams and try to navigate to the North Pole. So far, in the more than
three hundred years of struggling to reach the North Pole (and by extension, a long-sought shorter route to China and the Indies), the so-called
"polar problem" had yet to be solved. Norway's renowned explorer Roald
Amundsen had recently become the first to navigate the Northwest Passage by boat in a three-year journey from 1903 to 1906, but the very top
of the world remained the Arctic grail still highly coveted by the world's
most intrepid explorers.

Wellman learned much in his first two attempts. North of Spitsbergen—
a large archipelago in the Arctic Ocean above Norway—in 1894, his hired

ship the *Ragnvald Jarl* was crushed by giant fangs of ice and sunk, and he and his fourteen crewmen—after reaching 81° N—retreated over the moving ice pack, dragging their boats and portaging their sleds over dangerous open leads of water. They finally straggled back onto land, gaunt and hypothermic, after two long months of ice travel. They were all rescued by a Norwegian sealing vessel and returned to mainland Norway. Of the ordeal and defeat, Wellman penned a dispatch to the *Evening Sun* titled appropriately, "The Ice King Wins."

His second trip, to Franz Josef Land—a vast archipelago northeast of Spitsbergen, above Russia—in 1898, was even more harrowing. After chartering the ice steamer *Frithjof* and sailing from Tromsø, Norway, Wellman—along with three other Americans and four Norwegians—set up a base camp at Cape Tegetthoff on Hall Island and an advanced base camp forty miles farther north, dubbed Fort McKinley after the U.S. President, and Wellman's supporter. Of the two men left to winter there in near-total darkness in a crude, stone-walled hut, one died of unknown causes, and the other barely survived with his sanity after sleeping in the hut with his dead companion for nearly two months. When Wellman and his small team arrived at Fort McKinley in the spring of 1899 for his polar bid, he found the survivor, Paul Bjørvig, despondent, soot-covered, and shaking with fear. Wellman discovered the frozen body of Bernt Bentsen inside the hut. Shocked, he asked Bjørvig why he hadn't buried him.

"Because I promised him I wouldn't," replied Bjørvig. "So the bears and foxes wouldn't get him." Wellman and the others gave Bentsen a proper burial in a shallow grave and heaped stones over the body like a cairn, and Bjørvig made a cross out of scavenged boat thwarts, marking the date of his death: B. BENTZEN, DoD 2–1, '99. Then he spent hours in the numbing cold filling all the gaps between the burial stones, to ensure his friend would go unmolested by polar bears and Arctic foxes.

Wellman and his contingent—which now included the shattered Bjørvig—set out on a proposed seven-hundred-mile trek for the North Pole, but while sledging, Wellman had fractured his leg in an ice crevasse. He kept plodding north, but then, while camped on the ice, a violent storm struck, shattering the floe they were on. In the tumult of roiling ice, they lost many of their dogs and provisions and were forced to retreat. Wellman's leg was so badly injured that he had to be carried on a sled, and after many weeks of arduous slogging across the ice, they straggled into the camp at Cape Tegetthoff. When they cut Wellman's frozen trousers off, his leg was frostbitten and gangrenous. There was talk of amputation, but Wellman wouldn't hear of it. For the next four months he remained recovering in the house, immobilized, "tormented with the most agonizing itching; weak, feverish, despondent, sleepless."

Despite these trials and near-death experiences, Wellman remained, as he put it, "Under the influence of the Arctic spell. Its glamour was in our eyes, its fever in our blood." He vowed to return, only this time, he would employ the most advanced technological means humanly possible. He had already made the bold prediction, in a newspaper article titled "Walter Wellman on Future Modes of Travel," that "Within 25 years aerial navigation will solve the mystery of the North Pole and the frozen ocean." His harrowing experiences with traditional means had solidified that belief.

There were plenty of doubters who thought his ideas and predictions far-fetched at a minimum, and some considered them beyond possibility and even insane. Of the men (and they were all men at the time) vying for the top of the world during the late nineteenth and early twentieth centuries, Walter Wellman remains the least known, and in many ways the most unlikely. He could also be called the most visionary, a controversial claim. But Wellman rarely shied away from controversies; he sometimes intentionally created them. That was part of the newsman in him, which began at a very early age.

Born in Mentor, Ohio, in 1858, Walter Wellman was a pioneer from his boyhood. His parents (and his four brothers, of which Walter was youngest) moved around, spending five years in the Michigan woods before they struck for the western frontier, settling in 1871 in a small, squat home dug out of a hillside near York, Nebraska. Walter left home at the age of twelve and worked as a country store clerk for a couple of years until, with practically no formal schooling, he created his own newspaper, the *Sutton*

Times, in the fledgling town of Sutton, Nebraska, at the age of just four-teen. It was the town's first newspaper.

By his early twenties, he was back in Cincinnati, Ohio, founding *The Penny Paper* with his brother Frank. Walter and Frank eventually sold *The Penny Paper* to the Scripps Brothers (James E. and Edward W.), who renamed it *The Cincinnati Post* and continued to publish it for the next hundred years. Walter Wellman married eighteen-year-old Laura McCann in 1879 when he was just twenty-one, and he now had some capital from the sale of his paper, which he would need to raise his grow-ing family of what would eventually be five daughters (Ruth, Rose, Rae, Rita, and Rebecca, in relatively rapid succession). Intense and indefatiga-ble, Wellman would spend the next decade establishing himself as one of the most prolific correspondents in the United States, writing hundreds of pieces chronicling economic calamity and prosperity, American progress in transportation and communication (the railroads and the telegraph sys-tem), and most important for his future in exploration, interviewing and writing profiles about captains of industry and leading political figures of the day, including a major coup when he interviewed President-elect Gro-ver Cleveland just ahead of his March 1893 inauguration.

But even with that solid scoop, Wellman's attentions were already lead-ing elsewhere, as were the attentions of his readers. By the end of the nineteenth century, polar mania held everyone, including Wellman, in an icy grip. For decades the world had been enthralled by the fate of the lost Franklin Expedition (1845). Sailing in the HMS *Erebus* and HMS *Terror* looking for the Northwest Passage, Sir John Franklin and 129 men, all of the officers and crew, were destined to die horrible deaths in the Arctic by starvation, exposure, pneumonia, and even lead poisoning after their ships were caught icebound. Scores of relief expeditions went searching for clues about Franklin and his men, often encountering disasters of their own, and the reading public was gripped by the salacious tales of mutiny, murder, and misery on the icy seas.

For many Americans, however, including Wellman, it was the fate of the Lady Franklin Bay Expedition (aka The Greely Expedition of 1881–84) that brought polar exploration into their homes and consciousness. Under the auspices of the U.S. Army, Lieutenant Adolphus W. Greely and two dozen men went by ship to Lady Franklin Bay, near the northernmost tip of eastern Ellesmere Island, Canada, where they built a fort and, as part of a twelve-nation global consortium of bases circumnavigating the polar region, were to perform scientific research for three years. The plan was for the U.S. government to send resupply ships each summer. But a combination of poor ice conditions, bad planning and decision-making, and terrible luck prevented the promised resupply ships from reaching Fort Conger, and after two winters there, Greely made a fateful decision: he would take all the men in a twenty-eight-foot-long steam launch and a few whaleboats and head south 250 miles to Canada's Cape Sabine, where prior orders had stipulated the government would cache food and provisions as an alternate plan should they be unable to reach him. Shipwreck subsequently prevented the government from depositing the promised food.

What followed was a terrible ordeal among a labyrinth of massive ice floes, some up to fifteen miles in length. After being forced to abandon the cumbersome steam launch, Greely and his men floated on a small ice floe for over a month before finally making it back to land in the vicinity of Cape Sabine in October of 1883, where they built a crude stone shelter just as the darkness of polar night descended on them. Here, Greely and his men endured unimaginable hardship. Supplemented by just a few seals and a single polar bear, Greely rationed two-months' food to last eight months. They were reduced in the end to eating lichen and scurvy grasses foraged from the wind-pummeled coastline, and Greely had one man executed by firing squad for stealing food from the scant remaining rations. When a naval relief ship at last arrived in June 1884, rescuers found a macabre and heartbreaking scene: a dilapidated and shredded tent blown to the ground; corpses of men half buried; and just seven of the original number still alive. One of the unfortunates, discovered with amputated legs and a spoon lashed to his frostbitten hand, died during the return.

When they finally reached home, Greely told of their achievements: despite their hardships, thousands and thousands of scientific measurements were taken—of wind speeds and temperatures and ice thickness and sea currents; reports and records of hundreds of miles of previously unexplored lands on Ellesmere Island and along the coast of Greenland; hundreds of photographic plates of the newly explored and charted regions; and last, the report that two of his men, David Brainard and James B. Lockwood, had set a new Farthest North record in May of 1882, reaching 83° 24' N, just 396 miles from the pole. It was the first time in three hundred years that a non-British expedition had achieved Farthest North.

But the accomplishments of Greely's expedition were indelibly marred by the brutal truth that only six of his men came back alive, and worse, word soon reached the press—leaked by one of the relief sailors—that some of the bodies disinterred at Cape Sabine and brought back to the United States had been stripped of their flesh, likely the result of cannibalism. Immediately, *The New York Times* ran a story with the following headline:

"Horrors of Cape Sabine—*Terrible story of Greely's Dreary Camp—Brave Men, Crazed by Starvation and Bitter Cold, Feeding on the Dead Bodies of their Comrades . . .*"

The headlines and rumors plagued Lieutenant Greely for years, though he went on to have an illustrious career and became a founding member, and first president, of the Explorers Club and a founder of the National Geographic Society.

Walter Wellman was as well-read about Arctic exploration as anyone alive. He had followed Greely's story closely. Wellman had also tried to reach the North Pole the way all others had always tried, with the same result. When the relief ship had come to retrieve his expedition at Cape Tegetthoff in Franz Josef Land in late July of 1899, Wellman, now walking with a hand-hewn cane, limped down to the beach, thankful to be alive. He was going home after more than a year in the Arctic. He surveyed the rugged scene: the two prominent spires thrusting from the water off the headland, the vast plateaus of rock and ice blending into the horizon. He watched as graceful Arctic terns flitted and dove, then hovered effortlessly on the wind. At that moment, Wellman believed one thing with clarity: If he or anyone else was ever going to reach the North Pole, they would need to fly.

2

THE WELLMAN *CHICAGO RECORD-HERALD* POLAR EXPEDITION

"GO FIND THE POLE" Startling Assignment to a Washington Correspondent—WILL UNDERTAKE TASK—*Evening Star,* Dec 30, 1905

THE HEADLINES IN THE NATIONAL NEWSPAPERS OF THE UNITED States in the last days of 1905 and the first days of 1906 announced that the inimitable Walter Wellman—whom they had reported a half decade before "Wellman Back, a Cripple"—would soon embark on another brave and perilous mission. Now, splashed across front pages from Manhattan to Los Angeles were stories of this bold new endeavor, in which the editor of the *Chicago Record-Herald,* Frank Noyes, had given his star reporter an important mission: "Build an airship, go find the North Pole and report by wireless telegraphy and submarine cables the progress of your efforts." The proposed airship, to be built in France, would (the editors claimed) be the largest airship ever manufactured. It was the kind of story that sent newspapers flying from newsstands and had paperboys hollering "Extra, extra, read all about it!" as they sold the evening extras on the streets.

Wellman had been dreaming about this idea, in various iterations, since returning home in 1900 from his disastrous second polar voyage. He had arrived back in the United States badly injured. His fractured leg took nearly two years of recovery, and he now walked with a permanent limp and a cane. But his leg wasn't the only fracture needing time to heal. "Two years were needed for the recovery of my health," he wrote of his homecoming, "and a longer time to pay off, out of my earnings as a journalist and writer, the indebtedness I had incurred in the Franz Josef Land trip."

By writing hundreds of articles, on subjects covering most of the central issues of the day during the first five years of the twentieth century, Wellman repaid all his debts and was finally solvent again. He rolled up his sleeves and pecked endlessly at his typewriter, generating articles about organized labor and union strikes, coal barons, and issues with the railroad. He traveled to Paris to cover an international balloon race, noting with great interest that the winning balloon traveled nearly twelve hundred miles in just thirty-six hours—enough, with the right wind, to take men to the North Pole. Column inch by column inch, he had managed to claw his way back to fiscal solvency—even if he had not yet been able to parlay the stories of his polar travels into a real wealth.

Wellman had other complications beyond financial ones to smooth out on his return to the United States in October of 1900. One matter was his marriage. While convalescing at Harmsworth House in Franz Josef Land the year before, he had spent ample time pondering over certain transgressions he knew he must remedy once he got home. Writing privately to his brother Arthur, he intimated deep regret for his prior actions: "I feel that I have been born again up here," he wrote, "I have too much to live for, and I want to get home and make my dear wife happy. I never loved her as much as I do now." He went on to say, "If the fates spare me to get back to her I will make full amends for all my wrong-doing." Arthur knew exactly what his brother was referring to: for the past couple of years, Walter Wellman had been having an extramarital affair with a young woman named Miss Willard, and she had given birth to a child. Laura Wellman knew about the affair too.

Laura Wellman forgave her husband—at least temporarily. She had five girls to care for and a house to run, and depended upon him as a provider, so staying with him was as much out of necessity as it was out of love. As for Miss Willard and the child, the less Laura knew about them the better. Wellman had secretly corresponded with Arthur concerning his mistress and illegitimate child, explaining that the affair was over:

> As to Miss Willard, who last I knew was living in Hanover, Germany, my wishes are these—She shall be treated justly, generously. I left an assignment of $3500 life insurance to her, for the child's benefit . . . I shall provide for the child and help her as a man of justice would treat a divorced wife, but I should *never* go to her, or be with her, or even see her if I could avoid it . . . I respect her and admire her . . . She is a magnificent, noble girl and I wish her happiness and contentment. But my *love* henceforth is all for my wife . . . I shall attend to Miss W's future, so far as I have anything to do with it, and that is wholly financial . . .

With that, Wellman effectively compartmentalized his indiscretion, and he and Laura moved forward with their life together.

His wife Laura's trust was not the only trust he needed to win back. Wellman had always planned a return to the north, but such ventures were extraordinarily expensive. Paying off creditors helped, but convincing them to back future risky endeavors remained a challenge. Although he had not achieved his aims during his initial explorations, he had named a dozen or so islands for his patrons, and he only had one death on his conscience. Given the high mortality rate on polar expeditions, his record so far—at least in terms of getting men home alive—was well above average.

So, Wellman bided his time, repaired relationships, and waited for the right moment. Always enamored by the newest technology, he paid careful attention to all inventions and developments—especially those in transportation—and thought about how those might be employed to improve polar travel over the use of the "primitive sledge." "It did seem that some such better means should be found," he wrote, "than the barbaric employment of sheer brute strength." To find the best method, Wellman "determined to watch the progress of the arts and mechanics to see if some better means . . . could not be found for advancing upon the Pole."

Those "better means" were being constructed and tested in France. On November 12, 1903, brothers Paul and Pierre Lebaudy, in a 185-foot-long motorized dirigible, flew from Moisson to Paris, a distance of thirty-eight miles. At nearly the same time, by comparison, Orville and Wilbur Wright were conducting their first test flights of the *Wright Flyer* at Kitty Hawk, North Carolina, but their heavier-than-air biplane was then managing to fly only two hundred feet. Wellman was deeply impressed by the news of the first functional motorized airship and paid careful attention to its development and progress as the *Lebaudy*—as the dirigible was called—was used by the French military to perform reconnaissance missions over the German border capturing aerial reconnaissance photographs from nearly five thousand feet, and even to drop dummy bombs to test its range and capabilities.

Then, while covering the Portsmouth Peace Conference in August of 1905, which took place in Portsmouth, New Hampshire, and was brokered in part by President Theodore Roosevelt, Wellman learned that the *Lebaudy* had recently flown sixty miles against a strong headwind, and that the ship had a lifting capacity of 7,500 pounds. Wellman posted dispatches about the conference and the subsequent Treaty of Portsmouth, which formally ended the Russo-Japanese War of 1904–05, but his mind was on the *Lebaudy,* and in the skies above the frozen north. "It at once occurred to me that if this type of airship was good enough for the purposes of war, it ought to be good enough for geographic exploration, as an instrument with which to extend man's knowledge of the earth . . . an airship adapted to the Arctic regions . . . I had never lost sight of the air as a royal road, where there were no ice hummocks, no leads of open water, no obstacles to rapid progress."

The splashy announcement of Wellman's new project, published worldwide at the end of 1905, came after months of behind-the-scenes planning, meetings, and high-level negotiations. Wellman, who by now was so close with President Roosevelt that the Rough Rider sometimes invited him to dinner "to talk about polar bears," sought the endorsement of the avid outdoorsman and adventurer. Roosevelt was intrigued, and later invited Wellman's employer Victor Lawson, publisher and owner of the *Chicago Record-Herald,* for a meeting to discuss Wellman's new and daring polar proposal.

Wellman had been meeting with Victor Lawson too. Lawson was a powerful and influential man whose family had grown wealthy in Chicago real estate. They also were part owners of a Norwegian-language newspaper published in Chicago called the *Skandinaven,* so Lawson had a personal interest in the people and regions of the far north. He also had deep pockets and an eye for a modern story. Wellman and Lawson discussed not only the proposed airship, which would be built in France, but also Wellman's idea to construct wireless telegraph stations in the Arctic and send instantaneous dispatches of the expedition, which would be a first, and a major publishing coup. In the cutthroat world of newspaper publishing at the time, being first to a story was paramount, often even more important than accuracy.

Drawing on his charisma and persuasive powers, Wellman managed to convince Lawson to pledge $75,000. But he still needed more. The airship itself would be very expensive, and the entire enterprise would cost well over $250,000 (nearly $8 million in today's currency). For the rest, Wellman wanted to attach scientific gravitas to the undertaking, so he sought support from telephone inventor Alexander Graham Bell and the National Geographic Society. The three-person board of directors (including Bell) convened and unanimously approved Wellman's proposal, resolving that "it is the sense of the board that the plans outlined by Mr. Wellman for reaching the North Pole are carefully and thoroughly considered, and give good promise of success; and that the board heartily approves these plans and will do everything it its power to aid in carrying them out."

One provision put forth by the National Geographic Society was that Wellman bring Major Henry E. Hersey along as the lead scientific representative. Hersey had served with Roosevelt's Rough Riders in the Spanish-American War and was Inspector of the U.S. Weather Bureau. Wellman happily agreed, and appointed Hersey as expedition meteorologist.

So, by the end of 1905, after having corresponded with French airship designer Louis Godard for initial planning, and after securing a contract with the American De Forest Wireless Telegraph Company so that he could report his adventures to the world, Wellman was ready to embark on his next grand adventure, the Wellman *Chicago Record-Herald* Polar Expedition. Wellman conceded that the title was "dreadfully long and clumsy," but he wanted his own name included, as well as that of his newspaper, which was the main financial sponsor.

Now, all that was left for Wellman to do was explain to his wife, Laura, and his daughters that he was leaving for Spitsbergen again for at least a year.

That, and build one of the largest airships in the world.[1]

---

1 The French semi-rigid *Lebaudy III* was 185 feet long. Count Ferdinand von Zeppelin's experimental *LZ1* and *LZ2* rigid airships were both over 400 feet long, but as of 1905, the *LZ1* had been decommissioned and sold for parts and the *LZ2* had recently crash-landed and been demolished.

3

THE *AMERICA*

IN WHAT WAS A CONCESSION TO HIS WIFE, LAURA, AND HIS FAMILY, Walter Wellman arrived in Paris by luxury steamer in January 1906 with two of his talented daughters: Rose (22) and Rae (20). The two young women were popular in society circles, and Wellman agreed it would be exciting for them to explore Paris while he got to work on his airship.

Wellman had faced plenty of reaction in the press after his newspaper announced the "Startling Assignment to Washington Correspondent—." Some headlines were positive, even laudatory, including "Praise for the Pole Hunt" and "Another Dash for the Pole." One profile on Wellman said, "It is a daring audacious thing he is preparing to do."

But he garnered a great deal of skepticism too. "Many declared the project was only a fake, a bluff," Wellman wrote of the reaction. He was criticized for doing it simply to gain notoriety for himself and advertising for his newspaper. While these would certainly be by-products, they weren't the main reason he was going, and it was no bluff. Wellman's limp and use of a cane was proof enough of the Arctic's harsh moods, and yet he was returning. Of the criticism, Wellman was circumspect: "Of course it was useless to reply . . . One could only go ahead with his work and do the best he could."

So that's what he did in the frenetic early months of 1906. He sought the advice of Alberto Santos-Dumont, the experienced Brazilian aeronaut. Santos-Dumont agreed to serve as a consultant and provide his technical expertise. He met with Henri Juillot, builder of the *Lebaudy Patrie*—at the time the most advanced airship in the world—to discuss design features, materials to be used, and the issue of "gas tightness," necessary to avoid the leaking of hydrogen gas.

Wellman also visited Louis Godard at his hangars and headquarters outside the city to talk about the airship's "lift" and payload specifications, potential range, speed, and to negotiate a contract for the airship's construction. Ultimately, Wellman was impressed by Godard, especially his flying experience; the man had by then made over five hundred ascents in balloons and prototype airships. Wellman chose Godard to build his craft.

They settled on an airship 164 feet long, 52.5 feet in diameter, with an envelope or "gas bag" whose volume would hold 224,000 cubic feet of hydrogen. The fabric or "skin" of the airship was crucial. Balloons had traditionally used varnished silk, but after making calculations of the Arctic weather and wind conditions with their unique pressures and strains, Wellman and Godard decided to add a newer fabric manufacturing technique: two layers of cotton fabric material and one of silk overlaid with three coatings of rubber to shed rain and snow, with every seam reinforced. The airship would be powered by two motors (one 55 horsepower, one 25 horsepower) driving two propellers, making it capable of producing speeds of 12–15 miles per hour. It would have a lifting capability of 16,000 pounds, which would be necessary given the weight of the enclosed fifty-two-and-a-half-foot-long gondola or "car," suspended below the envelope, the motors, fuel for the motors, three to five crew members, food, scientific instruments, and various equipment. It would be built by a Frenchman, but Wellman would name it *America*.

While Godard moved ahead with the airship construction, Wellman organized many other vital details. He intended to build a permanent camp at Virgo Harbor on Danes Island, Spitsbergen, which would be like a remote self-sufficient mining camp—though one set up to attempt something unprecedented in the annals of exploration and science. *The Illustrated London News* referred to it as "Mr. Wellman's scientific village in the Arctics." By ship (he had once again hired the *Frithjof*), they would take lumber, tools, and equipment to build houses to accommodate fifty or more crew and workers; outbuildings for fabrication and storage, including a machine shop; and, most important, a massive wooden hangar to house the airship, which would also be shipped there in parts and reassembled on-site. It was a staggering undertaking, and Wellman knew that time was short to get all of this done if he hoped to launch in 1906 as planned.

Wellman put Major Hersey in charge of getting the materials to base camp on Danes Island and to oversee construction of the buildings and hangar. To assist Hersey—and to eventually navigate the airship— Wellman had hired a twenty-six-year-old merchant marine officer named Felix Riesenberg. Riesenberg had been in Chicago when Wellman's new polar scheme was announced. He read about it in the newspaper and thought it was "One of the craziest enterprises ever thought of by man . . . crazy enough to seem workable." He went directly to the offices of the *Chicago Record-Herald,* where he found Wellman's editor, Frank Noyes. Noyes put Riesenberg in touch with Wellman, and the two men met. Wellman was impressed with Riesenberg's pluck, and by his credentials in the merchant marines. Wellman asked for his résumé in writing, and told Riesenberg he would get back to him. Not long afterward, Riesenberg received the following telegram:

REPORT AT TROMSØ NORWAY JUNE TWENTIETH– LETTER FOLLOWS Wellman

Riesenberg was happily stunned to receive the cable correspondence. "That wire," he wrote of the correspondence, "was a switch that shunted me into new adventures . . . led me toward the Pole . . . gave me almost everything—the bad with the good—shed fortune upon me and snatched it away, and in the end lifted me with happiness and depressed me with misery." Based on that one meeting, Wellman had hired Riesenberg as navigator, though the young man had never been in an airship, much less flown one. But then again, neither had Wellman, and that hadn't stopped him from pursuing his dream.

So Major Hersey, Felix Riesenberg, and Paul Bjørvig would convene in Tromsø in June of 1906 and, from there, convey everything by ship to Danes Island. Bjørvig held no ill will toward Wellman after his dark nights in Franz Josef Land with Bernt Bentsen and had signed on as leader of the sailing crew. Wellman would remain in Paris testing the airship motor, then steam from Paris with fifty tons of airship-related materials and meet the returning *Frithjof* at Tromsø to take him to Spitsbergen. En route, Wellman planned to stop along the Russian coast to pick up Siberian sled dogs to be brought on board for an emergency escape in the all-too-plausible event that the airship crashed.

Unfortunately for Wellman, delays by his airship builder Louis God-ard tightened his timeline significantly. Wellman had claimed he would be conducting test flights for all to see in Paris, but these never materialized. Godard blamed local labor strikes for the delays, but whatever the case, as June neared and Wellman was finalizing details for his departure, Godard "was unable to finish the mechanical part of the ship in time to permit the motor trials which had been agreed upon." At this point, with June approaching and Wellman aware of the short Arctic summer, he decided to purchase a Clément 50/60-horsepower motor to use as a backup—and take it, and the completed airship and car, to Danes Island, where he would assemble it all and test it there with his own mechanics. It was hardly ideal, and already Wellman was privately worried that his much promoted flight would never leave the ground in 1906. But he would do everything in his power to get *America* airborne and headed north.

With his typical hustle and can-do approach, Wellman chartered a special freight train to transport the one hundred thousand pounds of airship parts and equipment from Paris to Antwerp, Belgium, where the steamship *Frigga* would meet him. From there, the *Frigga* would travel a thousand miles to Tromsø, at which point *Frithjof* would make two or three trips shuttling loads to Danes Island. The logistics were overwhelming, but Wellman ticked them off one task at a time.

By early June, Wellman had bid farewell to his daughters Rose and Rae, and he stood on the railway platform in Paris, waving to a cheering crowd as they tossed flowers and called out *"bon chance!"* Wellman doffed his hat, boarded his car, and the train chuffed out of the station, heading north.

In Tromsø, Major Hersey and Felix Riesenberg were reeling drunk. So was Captain Nordby, charged with sailing the *Frithjof* to Spitsbergen. They had spent the last few days loading the *Frithjof* with baggage and stores, boxes of gear and tools, hydrogen gas apparatus, tinned meat, and tons of malted milk products. "We wedged and shoved and filled every cubic inch of the hold," reported Riesenberg. But there remained piles of gear on the quay that still had to be loaded, and tons of lumber for the hangar and houses to be built remained on a warehouse floor adjacent to the quay. The craft was bursting to the main hatch, so they covered and battened it down with layers of tarpaulins, trying to figure out where they were going to put the much-needed lumber.

Each of the last two days, the Tromsø port wardens had visited the *Frithjof* at the dock and voiced concerns about overloading, saying that the ship rode "dangerously low already." On their visit that day at noon, they had informed Hersey and Riesenberg that no more could be loaded or the ship would not be permitted to leave port.

In the afternoon, Riesenberg had an idea. He invited the port wardens for drinks at the nearby Tromsø Club, a noted local watering hole where the town leaders, fishermen, and Arctic explorers often gathered. He bought a round of highball beakers of whiskey and soda, then another, and another. Riesenberg reminded the port wardens how much commerce Wellman's expedition was bringing to the town. By early evening, the inebriated wardens had softened, and when Riesenberg suggested that the lumber could be lashed down onto the decks with chains, and would "add to the *Frithjof's* buoyancy," they drunkenly agreed to the lumber, then added, "But not an ounce more of loading—not one ounce."

After an hour or two, the place was filled with people and raucous with laughter and storytelling. Time blurred. Hersey wobbled at the end of the table as Riesenberg toasted "The North Pole!!!" The port wardens raised their sloshing glasses and belting out "Skaal!" and toasted the Wellman *Chicago Record-Herald* Polar Expedition, slurring the clumsy title in their broken English.

At around 2:00 A.M. Riesenberg stumbled out of the tavern into the midnight sun. He shielded his eyes, then ran around the block three times to try to sober up. Then he hurried to find Paul Bjørvig. "Have everyone on board at once," he told him. Bjørvig hurried to assemble everyone on board: more than thirty hired men, including sailors, carpenters and laborers, sail-riggers, cooks. Riesenberg and Hersey ordered the rest of the gear and equipment loaded as well, lying that the port wardens had given permission for more than the lumber, which was removed from the warehouse and lashed down. Hired men hurried to load an additional twenty tons of sacked blacksmith coal on board, along with medical stores, more scientific instruments, and more tool chests "the size of coffins."

From *Frithjof's* deck, Riesenberg could see a group of revelers reeling out of the Tromsø Club. Among them were the port wardens and Captain Nordby. Nordby shook their hands, clapped their backs, and broke away, stumbling toward the *Frithjof.* A couple of straggling laborers were so drunk they had to be dragged on board and carried to their bunks. "Captain Nordby, the old Viking," wrote Riesenberg, "kept his pins and laboriously gained the little bridge." They got underway at once.

As *Frithjof* sailed away, the gunnels rode just a foot above the water. "We were awash fore and aft and worked the pumps constantly," said Riesenberg. As they cleared the docks, Riesenberg used binoculars and looked back, chuckling at the port wardens, who stood with their hands on their hips, dumbfounded and mouths agape as they stared at the completely emptied-out warehouse. Not a single ounce had been left behind.

4

CAMP WELLMAN

WHEN WELLMAN STEAMED INTO VIRGO BAY A COUPLE OF WEEKS after Major Hersey and Riesenberg, he saw progress. The *Frithjof* had made multiple trips back and forth from Tromsø, and his trusted men had not been idle, managing to unload the hundreds of tons of prefabricated lumber and metal needed for the houses, outbuildings, and hangar, as well as one hundred twenty-five tons of sulfuric acid, and according to Wellman, "thirty tons of apparatus and other chemicals for the manufacture of hydrogen gas." Additionally, using rowboats and improvised rafts, they had unloaded "the aeronautic machine and all its appurtenances: dog sledges, motor sledges, a steam boiler and engine, tons of gasoline, tools, coal, iron rods, bolts, nails, steel boats, and all the paraphernalia" they would need. They had worked tirelessly for weeks, and had even managed to frame out a large, double-walled house, lighted by skylight windows, and spacious enough to house forty scientific staff members. Major Hersey dubbed the place "Camp Wellman."

What Wellman did not see as he was taken by dinghy onto the stony shore was the massive hangar that he and his Swiss engineer Alexandre Liwentaal had carefully designed, the stupendous structure that was to house the airship *America* once inflated. The floor area needed to be two hundred feet long by eighty-two feet wide, with room for nine wooden arches, each over eight-five feet high.

The problem, Wellman learned once he was greeted by Major Hersey and Felix Riesenberg, had been finding a large and level enough area for the structure. The best site they had settled on, just inland from the Camp Wellman base headquarters, had in its center "a mighty rounded volcanic rock" which could only be removed by dynamite. No one there had ever blasted with dynamite before, so they were uncertain of the proper technique, and more important, the amount of explosives to use. Wellman walked around the area, considered the great monolith slowing down progress, and said to Riesenberg, "Take charge, Felix."

Although he was entirely untrained in the use of explosives, Riesenberg relished the responsibility and oversaw the drilling of three holes into the frozen ground directly beneath the offending rock, which was as large as a small house. With the drilling finished, there was the important matter of how much dynamite to use. No one seemed to have any idea. "Use at least a kilo of dynamite," Wellman finally said. How he arrived at that number was a mystery.

Riesenberg went to a storage shed and found cases of dynamite procured from the Nobel works in Sweden. The cases—stamped all over with danger symbols—were packed with "small sticks, tinted pink." A separate box contained the detonating charges and fuses. Riesenberg carefully brought a two-and-a-half-kilo carton of dynamite to the site and, fixing it with three detonating charges and measured-out fuses, slid it down into the largest drill hole. He then did the same at the remaining two holes. He and a few others then filled the holes with snow and ice. When all was ready, Wellman asked Riesenberg what amount of charge had been placed, and Riesenberg just shrugged: "Enough to crack the rock," he replied, then cautioned Wellman to have everyone move back at least five hundred feet.

With that, Riesenberg lit the fuses and sprinted for cover.

The explosion boomed across the valley, echoing like thunder against the mountainous hillsides. Parts of the boulder flew through the air, "their whirring jagged edges whistling," and one huge, fractured chunk landed like a meteor near the *Frithjof* and sent waves lashing against its hull. Birds screeched and flew from their cliff nests and the dogs set to howling.

When the percussive echoes ceased and it was quiet again, everyone hurried to gather around the hole left by the blast.

"Felix, you certainly lifted that rock," Wellman said.

With the rock cleared, Wellman and the men settled into a strict routine of fourteen-hour workdays. There was a tremendous amount to do. They finished the *Chicago Record-Herald* House, which Wellman described as "The best house ever put up in the far north," with "an outer corridor for stores, surrounded by double walls . . . thus the habitation is always warm and dry, even in the depth of the Arctic winter." The house included a bathroom with a porcelain tub.

Wellman supervised the construction of the numerous outbuildings and sheds, including the machine shop, which contained drills, lathes, saws, and other tools; the boiler house, steam pump, steam engine, and importantly, the building to house the gas device which would be used to inflate the *America*. Indefatigable, Wellman limped from one job and crew to the next, always busy and always encouraging, doing, according to Felix Riesenberg, "ten men's work in planning and carrying the . . . responsibility."

But the biggest and most complicated job of all was the building of the airship hangar. It took a week to lay the foundation and install the wooden floor. For this task, Wellman was able to scavenge timber from the remnants of S.A. Andrée's balloon hangar just a few hundred yards down the beach to the west, from which the Swedish balloonist and his two compatriots had drifted in the *Eagle* seeking the North Pole back in 1897. Wellman had searched for the lost explorers during his expedition to Franz Josef Land in 1898, hoping that by some miracle they had made it there and managed to survive—since the prevailing winds when they had last been seen might have blown them in that direction. At the time, Wellman understood that finding S.A. Andrée would also be a major journalistic story. But Wellman had found no sign of Andrée. "Great was our disappointment," Wellman wrote of the experience, "for we realized that thus ended all reasonable expectation that the brave Swedes were to be seen among the living."

The hangar was a complex engineering problem. The large, spanning arches—over which an acre of canvas would be stretched for the roof covering—had to be fabricated on-site, the lengths of lumber bolted together and then shaped on special bending frames. The massive but intricate arches were stayed with heavy wire and held taut with turnbuckles and hooks. As the arches were completed, they were raised using a pyramid-shaped derrick requiring twenty to thirty men. The original design had called for nine equally spaced arches, but given how long it was taking to build, shape, raise, and place them, Wellman decided to reduce the number to five to ensure a working hangar in which to inflate the airship and attempt a voyage. Wellman knew that fewer arches made the structure weaker and possibly subject to blowdown in high winds, but with summer drawing short, it seemed the best option, and even with only five arches the project took nine weeks to finish. When completed, it was by far the most impressive and conspicuous structure on Danes Island. "It rose there," wrote Riesenberg, who was proud and awed by the labors of the Norwegian carpenters, "magnificent, a . . . monument to our short summer of titanic labor."

Wellman had chosen Virgo Harbor on Danes Island on the Smeerenburg Sound for a variety of reasons, some practical, some symbolic and historically important, and some commercial, geared toward headlines. On the practical side, he knew that the great Gulf Stream terminated near northwest Spitsbergen, where the relatively warm water flowed through a deep channel before colliding with the icy Arctic waters and eventually meeting the frigid ocean current over the polar basin. As a result, even at its high latitude of nearly 80° N, the waters there remained warm enough to provide open, ice-free passage for vessels almost year-round.

Wellman had studied and written about the history of the place, which, beginning in the early seventeenth century, was a vibrant seasonal city developed primarily around whaling and walrus hunting, but also for sealing, polar bear hunting, and even the harvesting of down from the thousands of cliff-dwelling eider ducks. Dutch, Russian, Norwegian, and Scottish ships came by the hundreds each summer to exploit the resources, and the place was founded and named "Smeerenburg" by the Dutch in 1614—the term linked to their expression for rendering whale blubber and translated as "Blubbertown." At the height of the summer season, it swelled to a population of three to four thousand, a lawless place reeking of boiling whale blubber and viscera, with dance halls, cafés, bars, and brothels. The remnants of the blubber ovens—and graves of Dutch whalers—were just down the beach, between Wellman's base camp and Andree's dilapidated balloon hangar. Wellman appreciated the historical significance of the place.

More recently, ever since S.A. Andrée had used Danes Island as his base—and now because of Wellman's much-publicized North Pole plan—the fjord had become something of a tourist attraction, frequented by rich and adventurous sailing enthusiasts seeking a true Arctic experience. While Wellman oversaw work on the camp that summer of 1906, he was visited by the Prince Albert of Monaco, who had keen interest in marine science. "He came to our little port," wrote Wellman, "with his magnificent yacht, the *Princesse Alice,* and spent several days with us, having us out to dinner, and accepting the rude hospitality of our house in return." Wellman found Prince Albert to be delightful and sociable, and the two spent many hours together discussing Wellman's bold plan, science and the arts, as well as global matters of state, in which both men were well versed. Wellman was drawn to and comfortable with powerful and influential people, and they in turn seemed compelled by his charisma, so it was a natural comradeship.

When not working, Wellman also spent time conversing with the many journalists and even motion-picture company representatives who had come to document Wellman's historic polar flight. Among them was one journalist Wellman was acquainted with, an affable, multilingual German named Otto von Gottberg who wrote for the Berlin *Lokal Anzeiger*. Wellman had been seated at the same table with Gottberg at the

Portsmouth Peace Conference a year earlier, and the two had developed a rapport. Now, Gottberg was at Danes Island to write about Wellman's expedition—or so he claimed. Wellman's engineer Alexandre Liwentaal wasn't so sure that the German's motives were purely journalistic. Liwentaal, who had personally worked for Count Zeppelin conducting airship experiments in Germany and Switzerland,[1] told Wellman that Gottberg was a retired German military officer "who had been sent . . . by the German General Staff to watch our proceedings." In short, according to Wellman's engineer, the man was not a journalist but a spy.

Wellman scoffed at the notion and invited Gottberg to stay as long as he wished and to observe any and all construction and proceedings and assured him he would answer any questions he might have. There was plenty to see. Always with an eye on the ways in which technology might improve over-ice travel, Wellman had been scheming on a novel idea to replace the traditional dog and dogsled approach. Dogs were hard to control, and they needed to be fed, which in turn required carrying heavy loads. Wellman understood that should something go wrong and his airship went down, he would need to have with him either dogs or some alternative to use in continuing to the pole and back, or for retreating to safety.

His conceived solution was the so-called motor-sledge, and he had brought some to test out on Danes Island prior to the airship launch. They were prototypes of the modern snowmobile. The idea—a modification of the much earlier (and much heavier) "steam sleigh," invented in Illinois in 1836—was to employ gasoline-powered engines on a seated contraption with room, theoretically, to carry hundreds of pounds of gear. Prior to leaving Paris for Spitsbergen, Wellman had commissioned the design and

---

1 By the time Wellman was at Danes Island in the summer of 1906, Count Zeppelin had designed and test flown his airships *LZ1* and *LZ2*, but they were as yet not considered successful enough to be adopted as good investments for the German government. Though Wellman dismissed the idea that the Germans were spying on him for military advantage, it's worth noting that Zeppelin himself visited Danes Island in 1909 to consider it as a base for an airship flight of his own to the North Pole, though he never went through with the project.

building of a few different types, and had published articles, including in *The New York Times,* promoting his unprecedented use of them in the Arctic. They had been tested on frozen lakes on mainland Norway and showed promise.

Wellman had settled on a design that incorporated a large metal front wheel with V-shaped treads for traction. The empty wheel served double duty as a fuel tank. Behind the wheel, affixed to an axle, was an H-shaped sled platform and sled runners. Atop the sled platform were padded seats for two people—one driver and one passenger. The small combustion engine was situated beneath the driver's seat, from which the driver would steer with a T-bar resembling a bicycle handle bar. Otto von Gottberg and a few of his German contingent looked on with great interest as Felix Riesenberg assembled the mechanisms at the machine shop and then tested them.

Riesenberg was unimpressed by the motor-sledges. "They moved," he wrote, "and that was about it." Hampered by "weak little engines mounted on puny sledge runners," the machines crawled along slowly, sputtering and rattling, and as the fuel in the wheel ran low, the wheel bounced and pounded the hard ground as it went, leaking and then spilling gasoline all over the snow. "The terrible things were only one third as powerful as planned," mused Riesenberg, "and five times as heavy."

Wellman shook his head and cursed as the motor-sledges leaked and sputtered to a stop. Though they had been fine on the flat, smooth ice of a freshwater lake, they would be "useless upon the rough ice of the polar ocean," and he had them dismantled and packed back into their crates. While he was disappointed, he had a contingency plan in the form of a pack of thirty selected Siberian sled dogs, the best dozen of which he would now need to bring along aboard the *America* in the event of a crash or forced landing.

Despite the motor-sledge failure and the protracted work on the hangar, with teams of men working nearly round-the-clock, by late August Wellman had time to sit back, light a cigar, and survey their accomplishments:

"When all our work was finished we had a living house, a well-equipped machine shop, a balloon house, a hydrogen gas apparatus, a boiler house, a pumping house, and had upon the ground nearly two hundred tons of gas-making material."

He had also managed—assisted by representatives of the De Forest Wireless Telegraph Company—to set up a working wireless station at the headquarters at Camp Wellman, and one on the *Frithjof.* These could transmit messages to a station he had already installed at Hammerfest, on mainland Norway. His plan was to install another transmitter aboard the *America,* thereby bringing his Arctic expedition news to the world instantaneously. Up until that time, news from polar explorers had taken months—and sometimes years—to reach eager readers.

Wellman had followed Marconi's pioneering work in wireless telegraphy for the last decade, and was particularly impressed when, in 1901, Marconi successfully transmitted the first transatlantic wireless radio signal from England to Newfoundland. Though the signal was a simple "dot dot dot" Morse code for the letter S, even then Wellman knew that Marconi's achievement had the potential to revolutionize news reporting, when time was essential and outlets were racing to get the story to their readers first.

To illustrate his own pioneering use of Marconi's invention, that summer Wellman took a break from the work at hand and retired to the Camp Wellman main house. He rolled up his sleeves and sent a note which traveled magically from Danes Island to his wireless station at Hammerfest, then on to a station at Tromsø, which relayed it by transatlantic cable to the White House, where it was finally transmitted to Oyster Bay, New York. Sitting in his summer house, President Theodore Roosevelt had to smile as he held the message in his hands and read it: "Roosevelt, Washington. Greetings. Best wishes by first wireless message ever sent from Arctic regions. Wellman." The correspondence was picked up by the Associated Press and printed all over the nation.

Thrilled by the prospect of nearly instant correspondence with the outside world, Wellman immediately dashed off another message, this one to his own paper, the *Chicago Record-Herald*. The paper printed Wellman's entire dispatch:

WELLMAN FLIGHT SOON—IF WEATHER IS GOOD
Danes Island, Spitsbergen (via DeForest Wireless Telegraph to Hammerfest, Norway)
The big balloon house, in which the great airship will be sheltered, is now lifting its walls toward the sky.

The result is that we feel greatly encouraged by the rapidity with which the work goes on. If the weather remains favorable, as at present, we shall have an excellent chance at getting away for the pole this summer.

WELLMAN

Now came the biggest question of all. Would the untried *America* fly?

Wellman had great confidence in the airbag itself, which was as stout and well-built as any that had ever been made. The French balloon designer Louis Godard had a proven track record, and together they had agreed to add layers of fabric and rubber to ensure the minimum possible amount of hydrogen leakage. Wellman was less certain about the construction of the fifty-two-and-a-half-foot-long nacelle or "engine car" that would ride below the airbag and house the motors, the crew, and all the food and equipment—including tools and scientific instruments and boats, and the sleds and dogs for the journey. He had requested that the car be made entirely of steel, for maximum strength, but Godard had instead made it of wood, which was then sheathed with metal. It remained to be seen how the car would hold up once the motors were attached and running at full speed.

Before inflating the airship, Wellman and his engineers assembled the car on an elevated test block and affixed the motors near its center. The motors had been tested independently in France, but not while connected to the car and running the two propellers. The small group of engineers, workers, and a few journalists looked on expectantly as the engines were fueled up and started. At first, the motors hummed, and the propellers whirred beautifully, blowing paper and debris across the ground. But after a few hours, a cacophony arose, a disconcerting rattle and shaking as the propellers began twisting off their bearings.

"The driving gear went to pieces," admitted Wellman ruefully, "and the propellers could not stand even half of the strain which it was designed to put upon them."

Wellman, in disgust, ordered the engines cut. The staff and reporters backed away from the rickety machinery, giving Wellman space to fume. After he had spent some time venting and pondering his options, Riesenberg came to him with more bad news. The fuel for the motors was supposed to be transported in metal drums hung beneath the car in a large wicker basket, but "tests showed that the basket would fall apart if laden with the estimated amount of petrol" for the proposed one-hundred-hour flight.

The press jumped all over Wellman. Some called him a fake, suggesting the entire enterprise was a publicity stunt. But Wellman had thick skin, and he dismissed the criticism of what he called "the yellow journals." At any rate, he had always said, in interviews, his own articles, and even in his recent wireless messages, that he planned to try for the pole in 1906, but that the flight might be delayed until 1907. That was now looking like a reality. He would try again next year. The only good news was that the North Pole, ever elusive, yet remained unclaimed. He knew that Robert Peary, whom he considered his chief rival, was somewhere around northern Canada or Greenland, but so far, there had been no news of his expedition's results.

Forward-thinking and task-oriented, Wellman knew what he must do: Return to Paris, hire a new engineer, and completely redesign the *America*.

5

REBUILDING THE *AMERICA*

FOR WELLMAN, IT WAS A FRANTIC AUTUMN IN PARIS, 1906. HE HAD arrived in September, transporting his failed airship motors and the untested airbag by steamer. He had resolved "to enlarge and improve *America* for the campaign of 1907." The first thing he did was take the airbag to the cavernous Galerie des Machines at the former Universal Exposition building and inflate it to make sure it was undamaged and fully operational. The airbag appeared solid, but Wellman wondered whether it might be prudent to expand its length and volume by adding fabric extensions, which would increase its lifting capacity to nearly twenty thousand pounds. This would be necessary, Wellman decided, to accommodate his ideas for a larger redesigned car.

He had left Camp Wellman in the capable hands of Felix Riesenberg, whose job it was to keep everything in order there and maintain the many buildings, with the understanding that Wellman would return the following summer with a new and improved airship. Wellman had cautioned Riesenberg about the trials of Arctic winter, the so-called Long Night—months of near-total darkness. "You must recognize the dangers, the dangers to the mind, more than anything else," he'd told Riesenberg. "The depression of the dark, months of it, and no companionship other than a few men." Remarkably, the Norwegian Paul Bjørvig volunteered to remain as well, which was impressive considering his harrowing experience overwintering in Franz Josef Land in 1898, sleeping for two months beside the frozen body of his dead companion. Wellman was moved by the gesture, and before he left, he had given Riesenberg and Bjørvig a key to the liquor supplies. "Use anything you like, with discretion," he had said with a wink and a wry smile. "I trust you implicitly. You can use the Mullins' fifty-year-old rye, if you wish." Immediately after Wellman had left, they had tried to find the smooth aged rye, but could not, and settled for a lower-shelf bottle of Old Cluny.

In Paris that fall, it so happened that James Gordon Bennett, Jr., the owner of *The New York Herald,* was hosting his inaugural Gordon Bennett Cup balloon race. Bennett, an extravagant millionaire, society playboy, and sportsman, pioneered long-distance yacht racing, and, as he lived most of the time in Paris, had developed a fascination with ballooning and aeronautics as well. The winner of the "race" (it was not really about speed) was determined by the balloonist who flew the furthest distance from the launch site at the Jardin des Tuileries. Wellman was acquainted with Bennett and admired the flamboyant if eccentric man's journalistic wizardry. Bennett was best known for sending journalist-explorer Henry Morton Stanley up the Nile to find Dr. David Livingstone, who had gone missing in his search for the source of the Nile. Bennett had also sponsored his own North Pole bid, the ill-fated U.S. Arctic Expedition (aka *Jeannette* Expedition) of 1879–81, which resulted in the horrible deaths by starvation and exposure of twenty of the thirty-three crew. The terrible saga made for riveting reading and put Bennett at the forefront of the uneasy marriage of journalism and exploration.

The winner of Bennett's first balloon extravaganza was an American army lieutenant named Frank Purdy Lahm, who, along with Major Hersey—who had only recently overseen construction of Camp Wellman—flew from Paris to Scotland. After the race, Wellman sought out Lieutenant Lahm, and subsequently took balloon flying lessons from his father, F.S. Lahm, who was by then a fixture in the burgeoning experimental French aeronautics community. Wellman accompanied the elder Lahm on several thrilling balloon outings, including one that nearly ended in disaster. "We encountered a wind and rainstorm while over the

Seine," Wellman recalled, "and for a time it looked as if we were going to get a ducking in the river or be dashed against the buildings which lined its banks." Only by skillful navigation, and a great deal of luck, was Lahm able to bring the balloon down in a small yard between two buildings.

Wellman discussed various aspects of aeronautics with F.S. Lahm, as well as his immediate plans to retrofit his airship *America*. During their conversations, Lahm suggested Wellman meet with an energetic, courageous, and multitalented American named Melvin Vaniman, whom he believed might be useful to Wellman. Wellman immediately followed up on the introduction.

The man Wellman encountered was "of medium height, somewhat bald, his face lined, his features heavy, his eyes brooding . . . with a sense of revealing humor." As Wellman conversed with the vibrant forty-year-old, nicknamed "Van," he was impressed by his remarkably varied background. As a youth Vaniman had worked in mechanical shops and studied engineering. Vaniman was drawn to the arts as well and had studied and then taught music at Valparaiso University in Indiana in his late teens, and at just twenty he had joined an opera company and, for a decade, toured the United States as an opera singer. While touring with the opera company, he had picked up photography, specializing in panoramas and even designing his own cameras. This was all interesting, and though opera singing and amateur photography hardly qualified him to participate in Wellman's next airship expedition, elements of the man's subsequent career—especially his more recent undertakings—tantalized Wellman.

Seeking adventure and travel, Vaniman, by then married, had made his way to New Zealand and Australia, where he had found work as an aerial panoramic photographer for the Oceanic Steamship Company, which wanted sprawling bird's-eye images for their promotional materials. Vaniman became known for his daring feats climbing ship masts, tall buildings, special scaffolding, even a spindly one-hundred-foot pole swaying in the wind. He'd climb just about anything to get an image for his client. Eventually, wanting to go even higher, he had purchased his own balloon for this work, and traveled to Paris in 1904 to consult with the balloon and airship makers there. He and his wife, Ida, lived outside Paris, where, in a small workshop shed, Vaniman had started designing an airship he could use for his aerial photographic work. It was during this time designing and learning about airships that he had met the Lahms.

Inspired by Vaniman's courage, adventurousness, and industriousness, Wellman hired him immediately. "I was able to employ [him] as chief mechanic in rebuilding the airship," Wellman wrote of his decision. "Vaniman proved to be a splendid mechanic, and for the first time I felt that I could prepare designs and make plans with a reasonable degree of certainty that they would be executed."

Wellman gave Vaniman considerable autonomy, though he oversaw and consulted on all new designs. Vaniman got straight to work, redesigning Godard's car from scratch. He innovated by employing a V-shaped configuration of steel tubing and steel-reinforced wood and made it considerably larger at a hundred fifteen feet long and twelve feet wide, with an open top that would be covered in oiled silk to shelter the crew and gear from wind, rain, and snow. A twelve-thousand-gallon steel fuel tank would be attached below the new car and serve as a keel, a kind of "backbone or apex," the idea for which Vaniman said came to him in a dream. The fuel tank would be segmented with bulkheads to allow siphoning fuel from different sections to balance out the weight. For the motor, Wellman and Vaniman agreed on a single Lorraine-Dietrich 75-horsepower engine driving a pair of eleven-and-a-half-foot, two-bladed propellers, this time mounted on boom arms suspended from the sides of the car starboard and port rather than at the front and rear in the previous layout.

Working closely together during those early months of 1907, Wellman and Vaniman added a few more nuanced innovations. Vaniman, who liked to sing opera arias as he worked, created a clever platform that, when attached to the top of the tapered car, could carry up to six hundred pounds of food and be slid fore and aft, thus shifting the weight for balance and aiding performance. Vaniman also planned and began fabricating an enlarged rudder for greater steering control, and added removable planes designed to adjust the airship's vertical position. In addition, the clever mechanic included a solar sextant and drift indicators for navigational assistance.

Wellman integrated ideas of his own. To accommodate Vaniman's larger, heavier car—which now afforded, as Wellman put it, "a long and roomy cabin"—Wellman had the *America*'s gasbag cut in two, then spliced with twenty more feet of fabric, lengthening it from 165 to 185 feet and increasing its lift capability by two thousand pounds, to a theoretical lift force of 19,000 pounds. But his most significant addition was an altitude-controlling device he named the "equilibrator," though it was described variously as a "sausage," a "guide rope," and a "stuffed serpent." The idea—predicated on the assumption that maintaining a low altitude would reduce hydrogen loss—was to suspend from the airship a 120-foot-long leather dragline attached to the car by a long steel cable that could be spooled or unspooled to various lengths. The equilibrator was eight inches in diameter and covered or sheathed in thousands of overlaid steel scales for protection. In effect, the guide rope—borrowed from traditional non-motorized or free ballooning—served as "recoverable ballast." As Wellman explained it, "The guide rope is . . . an automatic control of upward and downward movement of the [airship]." To save space in the car, the tube of the equilibrator—divided into ten-foot sections, each one a closed compartment—would be stuffed with a thousand pounds of packaged foods for use in emergency.

Wellman's central concern, and his rationale for the equilibrator, was the concept of "pressure height." In a non-rigid airship such as the *America*—known as a "pressure airship"—the envelope or gasbag maintained its shape by keeping the hydrogen at a higher pressure than the outside air. The pressure height is the altitude at which the airship is one hundred percent full. At that point, if it rises to a higher altitude, hydrogen gas must be vented or valved to prevent the envelope from rupturing or even exploding, which would be catastrophic and likely fatal for the crew. With the heavy equilibrator dangling below and dragging along the surface of the ice pack, Wellman believed that the *America* would be prevented from exceeding pressure height. He would have no way of knowing for certain until testing it out in the Arctic.

An additional and related technology of Wellman's devising was the so-called "*retardateur*" (Wellman opted for French terminology in deference to the French aeronautical community). Somewhat like a ship's ice anchor, this was a long rope or cable whose end, as originally conceived, was covered with spikes to bite into the ice. Wellman had carefully studied the prevailing Arctic winds, particularly those of Fridtjof Nansen's logbooks from the three-year journey of the *Fram* (1893–96). Wellman knew that *America* would inevitably face strong headwinds that would overwhelm the engine's thrust. To keep them from being blown backward, he supposed that the retardateur, if deployed and working properly, would tether or anchor them in place until the winds subsided or became favorable again. But he conceded that the untried technology possessed uncertainties:

> It would be practicable to use an ice anchor and firmly anchor the airship to the ice floes, there to ride during head winds instead of drifting backwards on the course. But there was danger, with fast anchorage, that the strain on the ship might lead to breakage or accident. So, we compromised . . . and made the leather serpent covered with thousands of short, sharp steel points to scratch upon the surface of the ice as it was dragged along over the ice-fields . . . The equilibrator . . . with its smooth steel scales, was designed to make the smallest possible resistance; the [retardateur], on the other hand, with its steel scratchers, to make the greatest resistance its weight could effect as it was dragged along in contrary winds.

The new designs and fabrication took Wellman and Vaniman through the late spring of 1907, but by June, Wellman felt everything ready to transport back to Danes Island for another attempt. Once again, he was so behind on his schedule that he had not managed to test the completed airship in Paris. He rationalized—perhaps at his own peril—that tests in the atmospheric conditions of Europe would not simulate what the airship would experience in the Arctic, so he would test the *America* there. He brushed aside the doubts about his venture, which were numerous in the press and among the readership of the day, the criticisms ranging from "foolishly reckless" to "suicidal." Wellman dismissed his critics, suggesting that both the press and the general public was "always skeptical as to ventures and experiments which it little understands."

He hoped and trusted that his men had kept Camp Wellman in order during his absence, and that the hangar—which he had hurried them to finish with only five arches—had survived the winter. Despite his own uncertainty, he headed hopefully back north to Danes Island, once again aboard the *Frithjof.* But before leaving, he wrote an article for *McClure's Magazine,* which opened—with typical Wellman bravado—with something of a proclamation:

Some day in July or August of 1907 . . . A man standing at the northwestern point of Spitsbergen, six hundred miles almost directly north of the North Cape of Norway, will behold a strange and wonderful spectacle. He will see, rising from a little pocket of land amidst the snow-capped hills of Danes Island, an enormous airship—a huge mass of hydrogen gas imprisoned in a staunch reservoir of cloth and rubber, in shape much like a thick cigar, its sharp nose pointed northward.

Now, the pressure was on him to go and prove it.

6

WINDS OF CHANGE

FELIX RIESENBERG GREETED WALTER WELLMAN WITH A HAND-shake and a big-toothed grin when the expedition leader arrived back at Danes Island in late June of 1907. It had been a long year of waiting in the Arctic, but Riesenberg and Paul Bjørvig had survived in the heated confines of the Wellman House, while blizzards raged outside. They had spent the days before the sun departed trapping foxes, sometimes walking past "open graves holding the white bones of Dutchmen perfectly preserved." They had often paused to look at the bones of these whalers who had given their lives to this harsh place at the top of the world, noticing a large oblong stone set behind the ossified bodies "like a gigantic natural monument to the Arctic dead." They called it Sarcophagus Stone.

There had been one mishap, Riesenberg reported, gesturing toward a big scar, raised diagonally across his forehead. While out hunting, he had fired his rifle at a seal, and the gun's breech had exploded, knocking him unconscious onto the snow. He had awakened to find one of the sled dogs hovering over him, licking blood from his head. When Riesenberg sat up, he feared he had been blinded. He reached his frozen hand up to his face to assess the damage. "Ragged edges of flesh hung from my forehead, between my eyes," he reported. "My forehead was plastered with blood and snow." He had wiped his eyes of blood and discovered that he could see, but metal from the exploded breech had torn his forehead open to the bone. It was now healed but left him with a prominent scar, a permanent reminder to clear the snow from the muzzle of his gun.

During the long dark nights, the men at Virgo Harbor had sometimes gone out skiing in the extreme cold to take in the aurora borealis. The sky would burst into blood-red flames, then drip curtains of blue-black and green and yellow as the scene danced and changed and morphed into coronas. "Yellow petals," reported Riesenberg poetically, "millions of miles across . . . dipped down to the whited hills, their ends dropping bleeding beads. . . . Nothing man does with color, lights, or fire—his weak attempts at beauty—can approach this glory."

Wellman took in Riesenberg and Bjørvig's stories as Norwegian workers unloaded the *Frithjof* and brought the rebuilt *America* ashore. He was glad they had survived the winter, and impressed with the condition of the camp, which they'd kept tidy and in fine shape. He was relieved to see the hangar's framework still standing, knowing how strong the winter gales blew up there.

But gales blew in summer too. Not long after all the equipment had been offloaded and portaged to Camp Wellman and organized, a summer squall roared in ferociously from the north, shearing the guy wires and toppling the five arches. For the next month, men worked double-time repairing the hangar and erecting it again, this time with all nine arches per the original design. That put them into early August, and Wellman figured they had only a few weeks remaining to reasonably make their attempt. They needed to have everything ready immediately, and then study the winds, hoping for prevailing southerlies.

There remained the matter of assembling the crew that would risk their lives making the flight. One day at lunch, as Melvin Vaniman oversaw the configuration of the new 115-foot car, "a beautiful, tubular steel nacelle," Wellman took Riesenberg aside. "You are definitely chosen as one of the crew of the *America*," Wellman assured him. Riesenberg was deeply moved. To celebrate, Wellman pulled out a bottle wrapped in red tape, thanking Riesenberg for saving his fifty-year-old Mullins rye for over a year. When Riesenberg saw the bottle, he just laughed. "Saved it?" He chuckled. "Why sir, I hunted all over for it—in the stores, in the medicine chest, everywhere. We were ready, many times, to sink it."

Wellman held it up. "It's here, big as life."

"My God." Riesenberg laughed again. "I always thought that was a spare bottle of ink."

With that, sink it they did, taking shots all around.

Wellman would go, obviously. He had decided on a four-man crew, which would include Major Hersey, who was back again under the aegis of the U.S. Weather Bureau, and his new mechanical whiz, Melvin Vaniman. Vaniman had proved indispensable, and he now had the most intimate knowledge of the airship's construction and operations. In early August, Wellman ordered the airship envelope inflated in the repaired hangar. Manufacturing the hydrogen gas was a complex and time-consuming process that took six men, working twelve-hour shifts, twenty-four hours a day for a week. "Every day," wrote Wellman of the process, "they have to handle about ten tons of sulfuric acid, eight tons of iron shavings and scrap, pump forty tons of water, besides other chemicals for drying and purifying the gas . . . to produce the 280,000 cubic feet of hydrogen which the *America* holds." After being produced by the hydrogen apparatus, the gas was "washed" with water piped seven hundred fifty feet from a nearby glacier, then dried by being routed through a coke-filled cylinder, then removed of acid by piping it through a separate cylinder filled with caustic soda, resulting in hydrogen gas that was "nearly pure . . . very cool, dry, and free from acid." And potentially explosive. After inflating the envelope, they suspended the car below and tested the motor, and this time, everything appeared solid and operational.

Vaniman, who Riesenberg referred to as "a genius," tinkered with yet another innovation, this one designed to save precious fuel for the proposed hundred-twenty-plus hour journey. The device was a carburetor that could, with the flip of a switch, draw hydrogen from the gas bag via a silk tube about a foot in diameter into the Lorraine-Dietrich engine, simultaneously valving hydrogen as needed during ascent and reducing petrol consumption. But it was not without its dangers, including the potential of sparks from the gasoline engine igniting the incoming hydrogen. Wellman was dubious about it from the beginning, but when they tested the device, it worked. When Van hit the lever, the Lorraine-Dietrich, remembered Riesenberg, who had been looking on nervously, "seemed to spurt, to purr with added power . . . hydrogen, the perfect fuel gas, was burning in the cylinders."

Wellman responded by breaking out the remains of the fine Mullins fifty-year-old rye later that day.

"To Melvin Vaniman," Wellman said, raising his glass to the others. "He may shiver up here in the North, but he's not got cold feet."

Bad weather kept them grounded through August. Experienced sailors and fishermen cruising the waters around northwestern Spitsbergen told Wellman it was the windiest summer in thirty years. This weather system could either help or doom his chances. The summertime prevailing winds in that region, as well as over the 717 miles from Danes Island to the North Pole, fluctuate during the summer months. A period of strong southerlies might last a week to ten days, followed by a calming lull, and then a shift of equally strong northerlies for another week or ten days. Wellman had hoped the weather would cooperate and they would be able to perform some test flights, but time was too short now. Once they left the hangar and headed north, the pole attempt would be the test flight. He consulted with Riesenberg, and they agreed that September 5 would be the latest possible day they could depart and still get accurate observations of the sun once over the pole, so that they could determine their location and confirm they had reached the vaunted destination, the dream of explorers for three centuries. Saying you had been to the North Pole was one thing; proving it was another.

Wellman spent the last few days of August walking back and forth from the Wellman House to the hangar, then to the machine shop to the dog kennels and back again to the house. He checked and double checked that everything was ready, and there was much to inventory: ten months' worth of food and provisions; forty gallons of fresh water; 1,200 gallons of gasoline topped off in the tank below the keel; traditional sledges (he had scrapped the heavy useless motor sledges) and a small boat. *America,* now fully inflated, hung tethered and stupendous, floating in the Arctic air, and Wellman had much to be proud of. His dirigible was at that moment the second largest airship in the world, bested in size only by Count Zeppelin's *LZ3.* But Wellman was the one who was here, poised to make history.

As September 1907 arrived, the winds held steady from the north—precisely what they did not need. Everyone—especially Wellman—was on edge. The elaborate wireless system he'd paid a great deal of money for had not, in the end, worked properly, and the number of messages getting through was inconsistent. Wellman was incensed, having had to take over the operation himself when the representative from the De Forest Wireless Telegraph Company proved incompetent. To monitor wind conditions, Wellman ordered a sentry posted twenty-four hours a day on a nearby mountain peak. He was to survey the sound with binoculars and watch the *Frithjof,* which was steaming back and forth across Virgo Bay and would alert them of a desired wind shift the old-fashioned way: through semaphore communication from ship to shore, with a simple code displayed on the sails.

Then, Major Hersey informed Wellman that he was feeling ill and unsure he would be fit for the voyage. This added to Wellman's stress; Major Hersey was an expert meteorologist, skilled in taking readings and determining location, crucial for verifying that they had made it to the pole. Wellman had also counted on him to be their navigator. But Wellman adapted to the situation, quickly naming Riesenberg navigator and Vaniman engineer. Wellman would remain overall commander of the expedition. They would all be taxed to the extremes of their abilities, since there was no way to predict what challenges they would encounter in an untested experimental airship flying where no one had flown before.

Beyond the pressures created by the nearly endless tasks they all had to perform preparing for the flight, Wellman was also weighed down by the expectations and publicity he had created at home and abroad. He had made claims—some called them boasts—about the forthcoming journey of the *America,* printed in numerous national and international papers and journals, as well as in personal interviews he had granted to the global press. He had come to the Arctic three times before, each with the goal of reaching the North Pole. Supporters of the expedition had made large financial investments in the enterprise, and Wellman felt he must produce something this time, or he might not have another chance. And there was the matter of Robert Peary, who, last Wellman had heard, was back in the United States having work done on the ice-worthy steel-hulled *Roosevelt* and had predicted an August departure. Wellman knew he had a significant head start, but still the pressure to be first, and to get the story out, bore down on him.

Riesenberg also felt the deep responsibility of being named navigator at the last minute. He had been mentally prepared to assist, but now, with Major Hersey likely out, a great deal was riding on his abilities to steer the airship by compass, dead reckoning, and sextant when the position of the sun was discernable. Tension permeated all corners of the encampment. Someone had warned Wellman that the German officers in camp—who had offered to help tow *America* out into the sound for its launch, were spies, but Wellman dismissed the warnings as paranoia. Regardless, according to Riesenberg, "Our camp became a cage of shattered men, pacing about at all hours." There even arose a rumor that someone might try to sabotage the airship. Riesenberg attributed this to the general nervousness that had settled over Camp Wellman. When Wellman caught wind of the rumor, he ordered a round-the-clock guard at the hangar and holstered a loaded pistol to his belt.

"Let anyone try anything," he warned, "and I'll kill him."

At 4:00 A.M. on September 2, 1907, a signal flashed from the sentry on top of the mountain overlooking Smeerenburg Sound: "Calm to the north!" The winds around the camp swirled for a time, then quieted. The sled dogs yipped and howled, sensing a change in the weather. Wellman walked down to shore, alone in his thoughts. He was gone for a long time, staring out to sea, alternately watching clouds as they moved slowly across the mountains of Amsterdam Island and light haze lifting from the calm bay. At last, he turned and looked back at the Wellman House, at the magnificent hangar, at everything they had built over the last two years. He had been preparing for this moment, in one way or another, for over a decade. In some ways, he had been preparing for it his whole life.

He turned from Virgo Bay and strode back toward the hangar, his bad leg dragging slightly as he went to tell the others: It was time to go.

7

MADNESS

hangar for a private conversation. He wanted to be out of earshot of the staff for this talk. A nosy photographer saw them and sidled up with his camera, hoping for a candid shot on what promised to be an historic day, but Wellman shooed him away with a wave. The three men stood silent in the cold morning air. Major Hersey had been officially ruled out as a crewmember.

"I am determined to go," Wellman said at last. "How about you, Melvin?"

Vaniman didn't hesitate for an instant. "Of course. What else are we here for?"

Wellman looked to Riesenberg, who was shuffling in place. "And you, Felix?"

"It's what I've been waiting for," said Riesenberg flatly, betraying no sign of fear. "I'm ready, sir."

It was settled. Vaniman returned to the hangar to make last-minute preparations on the airship. He checked the two "ballonets"—internal air-bags inside the main envelope that could be inflated or deflated as pressure requirements dictated and to maintain the airship's shape. Riesenberg went to the dog run and, with a couple of assistants, brought the best of them back to the hangar and loaded the team in a specially made kennel at the front of the car. It took some coaxing and manhandling, but eventually, despite some balking and snarling and fighting, they managed to load ten dogs. Riesenberg had outwardly kept his nerves and doubts tamped down by staying busy, but inside, his stomach was roiling, his heart pounding. Now that it was really happening, the idea of casting into the Arctic skies above the polar ice fields, driven by a mere 75-horsepower engine, seemed ludicrous. "I was well scared," he admitted in his journal. "To [attempt] the bucking of polar winds . . . was madness . . . certain suicide."

An eerie calm settled over Virgo Harbor as the winds lulled to a whisper. The air was still and quiet, the sunrise a muted pink. "We'll go now," Wellman announced, arriving back at the hangar. Crewmembers drew back the heavy canvas hangar doors and tied them off. *America* was inflated fully and loaded for the journey to come. Vaniman was inside the airship car already, and Riesenberg climbed in next to him. Wellman, laboring on his bum leg, was assisted in last. The ground crew hustled to the hold-down ropes, grabbing on with their gloved hands. Men called out "Good luck!" and "Bon Voyage!" as Riesenberg settled himself at the wheel in the forward bridge. Vaniman dropped the last of the sandbags, then made his way to the engine room at the center of the car. *America* lifted slightly, beautifully balanced, bobbing effortlessly in the dimly lit hangar, with skeins of early light glinting off the rubberized envelope.

All was ready. "Walk her out," called Wellman.

Forty men began walking slowly forward, leading the craft out of the hangar and down the short, rocky beach to the water's edge, where *America* hovered some twenty feet in the air. Wellman ordered the airship swung while Riesenberg adjusted the compass and set the compensating magnets in their tray. "Compass ready," he called out, as men ran the towlines from the airship out to the *Express,* a little German steamer Wellman had enlisted to tow them away from shore for a safe takeoff. Wellman wanted to be clear of buildings and cliffs at the start in case of any sudden wind gusts.

"Slow speed," Wellman said, nodding to Vaniman, who started the engine. The *Express* eased forward at Wellman's signal, towing *America* past the sharp point of land just west of the camp and into the smooth water.

There was still very little wind in the bay, but beyond, to the north and west, clouds were massing over the mountains as they steamed slowly out into the two-mile-wide Smeerenburg Sound.

The airship hovered there, still tethered to the steamer, slowly gaining altitude. The men checked all the instruments and gauges: the thermograph to monitor the air temperature; the statoscope that measured falling and rising; the manometers that marked the pressure of the envelope and ballonets; the barograph that recorded atmospheric pressure. All were working optimally. When they reached an altitude of about five hundred feet, Wellman looked to Vaniman, then to Riesenberg, and then to the American flag fluttering at the stern.

"Cut the line," Wellman yelled from the forward cockpit. Riesenberg gave a double take to make sure. "Let go!" Wellman confirmed. Riesenberg followed the order, and as the towline spiraled down into the water beside the *Express,* they felt *America* lurch upward, then settle as the heavy equilibrator rose from the water as ballast, doing its job as designed.

The *America* was free, flying independently, the first airship ever seen above the Arctic. Over the percussive thrum of the six-cylinder Lorraine-Dietrich and the whirring of the propellers, they could hear distant cheers of the crew on the beach. As they flew forward into winds freshening from the northwest, Riesenberg looked back, and already Camp Wellman was shrinking in the distance, the great hangar appearing from their height as nothing but a small tent.

Vaniman pushed the engine to half speed as Wellman surveyed the crystalline landscape ahead. Rugged, snow-covered mountains ringed the sound to the east and west in a horseshoe, with the open end, due north, a cobalt-blue stretch of sea: the gateway to the ice pack of the Arctic Ocean, where they were heading. Suddenly, they heard an unnatural sound, a metallic clattering as a dangling towline caught in the port propeller, tangling in it. Vaniman quickly shut off the motor and shimmied through the side door and out onto the propeller strut. It was an acrobatic move, but his daring panoramic photography experience had prepared him for maneuvers at such heights, and he soon had the fouled line cleared and the engine started again.

"Full speed," bellowed Wellman once Vaniman was safely back at his post. Riesenberg adjusted the wheel to set their course due north. Looking down, the men could see growing riffles on the water, some surging to whitecaps, indicating strong winds bearing down from the northwest. Wellman cursed under his breath. Just a few minutes into their maiden flight, and already they were bucking headwinds, and the skies were darkening, dimming to a slate gray as ominous fogbanks rolled in from the west. They pushed on at what Riesenberg estimated was fifteen miles an hour.

Wellman had hoped and planned to clear the mountains in calm weather, but soon it began to snow, and though they could dimly make out—some thirty miles in the distance—the distinct line of polar ice, they all noticed the engine straining. "The beat of the struggling propellers sounded amid the roar of the exhaust," recalled Riesenberg. By checking land bearings—the two-thousand-foot peaks and sheer-faced cliffs now closing in all around them—they noticed their progress slowing, despite Vaniman driving the motor full tilt.

They were hovering at five hundred feet, with the equilibrator, "that long, leather, steel-studded snake filled with provisions," slithering through the sea like a dragon's tail. Wellman squinted from the cockpit, noticing snow accumulating on the envelope. Riesenberg worked hard to point *America*'s nose to the north, but the wind was driving the airship southeast, toward the looming mountains. Wellman knew that short of discarding the equilibrator, which he believed they needed to cross the polar sea, they had no way of quickly gaining enough altitude to clear the wall of cliffs and glaciers growing dangerously close just east of them. If the motor cut out now, or the propellers became fouled again, they would be dashed into the cliff walls and would almost certainly plummet to their deaths into the sea below.

Vaniman looked out the starboard window and saw they were in "imminent peril . . . drifting upon the high perpendicular mountain." He throttled up the Lorraine-Dietrich to its maximum output, and *America* just cleared the jagged outcrop as Riesenberg steered them through snow flurries into open skies. The winds abated, and Vaniman backed the motor down to half speed. For around twenty minutes, they enjoyed calm flying conditions, the airship gliding with lovely balance, forging slowly forward.

But the lull didn't last. Soon they were being driven backward to the south, gaining speed. Vaniman looked out the port window, alarmed to see that they were again careening toward a mountain appearing intermittently through mist, fog, clouds, and driving snow. They thought that the dull brown mass of rock must be Cloven Cliff, an island named for the deep split in its center. Riesenberg reacted and "tried to throw the ship's head to port; it answered slowly."

Riesenberg looked down wide-eyed as the compass needle spun wildly and erratically, its frenzied movement the result of some kind of magnetic disturbance. Vaniman clambered forward to see what was wrong. Riesenberg cranked the wheel hard, and they sailed past Cloven Cliff, nearly crashing into it. Gusts up to forty-five miles an hour then caught and spun the airship, whirling it in circles. Without the compass, and with visibility down to just a few hundred yards, they were effectively lost, uncertain of their precise location.

"Three times," wrote Wellman, "we came up so near the mountains, looming suddenly ahead out of the thick air, that we thought all was over, but each time the motor and propellers brought us round to temporary safety." But it was indeed only temporary. They had been aloft only a few hours, but everything was happening so fast it felt like minutes. Wellman knew that they had to act fast, and their options were limited. If they could somehow make it back to Smeerenburg Sound, they might safely be retrieved by either the *Express* or *Frithjof*, both last seen cruising the waters, watching them fly away. But in these winds, controlling the *America* had proved difficult.

A brief opening between the clouds revealed a large glacier, the blue-white tongue of ice carving through a narrow valley between two prominent peaks.

Wellman yelled that they must try to bring the airship down there. It was their only chance. They all watched as both the equilibrator and the retardateur, dragging and bouncing behind, slithered up a one-hundred-foot precipice at the foot of the glacier as they entered the valley. "They wound between and around giant rocks of the moraine," Wellman reported. "As we moved inland, the serpents fell into deep crevasses in the ice, then crept out again."

"Stop the motor!" Wellman yelled above the din of wind, whirling propellers, and engine noise. Vaniman cut the throttle and the propellers sputtered to a stop. They all met at the wheel, now powerless, a massive oblong balloon, "a mere toy of the winds," as Wellman put it.

"Bring her down on the glacier top," Riesenberg hollered, motioning to the mountain walls rising all around them. "Once back in those hills, we'll never get her out."

"Valve gas!" Wellman ordered. Riesenberg acted fast, pulling the emergency valve. Hydrogen gas whistled and squealed as it flowed from the envelope above them into the air. Descending, they saw the glacier rising up, deep fissured crevasses appearing like dark, bottomless canyons. Wellman released the anchor drag at the forward end of the car. It whooshed through the air and landed on the ice, and as it settled, the airship swung sideways, colliding, bouncing off, then ramming once more into a high wall of the glacier ice. Riesenberg shut off the emergency valve.

They remained aloft, moving fast and bouncing across the glacial surface. Riesenberg had shut the emergency valve just in time to avoid a catastrophic crash onto the ice. Vaniman scrambled to a wire connected to a razor-edged "crescent-like ripping knife" fixed to the inside of the envelope, designed to cut the air bag as a last resort.

"Rip, rip!" Wellman cried.

Vaniman yanked at the wire, but still they bounced along, the sound of wind and ice screaming as they caromed helplessly. Riesenberg, holding tight to the car rails, made it to the observation window and peered up into the dim insides of the sagging airbag. He could see the ripping knife hung up on a section of fabric. "Give her a jerk," he shouted, and Vaniman leaned back and yanked with all his weight, slashing a long gash in the fabric. There was a "belch of gas," Riesenberg said, "and the huge balloon above us collapsed . . ."

Thousands of square feet of fabric crumpled and puckered as the steel car struck the glacier surface and spun toward the edge of a crevasse.

8

"WHAT CAN BE DONE SHALL BE DONE."

ON FINAL IMPACT, RIESENBERG INSTINCTIVELY GRABBED ONTO A rope and was thrown somersaulting from the car. He skidded to a stop face-first on the snow and ice, his shoulders burning from the violent jerk as he halted his slide. He lay there for a time, wondering if he might have reached the afterlife. Then he could hear barking dogs, and at last Wellman's voice. Riesenberg opened his eyes.

"Felix! Are you all right?" Wellman asked, brushing snow from his clothes, face, and eyes. He had apparently also been thrown from the car. Riesenberg sat up, then stood. He was uninjured. Just then Vaniman appeared, crawling out of the engine room door. His face was blackened by grease and oil. "I got dropped into the saddle," he joked, pointing to the steaming engine.

The fabric of the airship envelope lay rumpled and folded in a heap beyond the car, and as they walked carefully around, they realized its acre of fabric spanned two crevasses. The car itself had come to a stop at the edge of "a gigantic crevasse, a hundred feet deep," according to Riesenberg. A few more feet and they would have all plunged into an abyss that, as he put it, "could have swallowed a dozen *Americas*."

Vaniman, always industrious and with a sense of humor, set a coffee pot to boil on the still-hot engine. Riesenberg went to check on the dogs, and found them all alive, though the wild ride had sickened some, and they had vomited all over their enclosure in the car. Consulting maps, charts, and what they could make out of their surroundings, they determined that they had crashed on the Svitjod Glacier, inland of Fowl Bay, about ten miles from Camp Wellman. They had flown for three to four hours, traveling a distance—including being blown back toward the south—of only about thirty-five miles.

They lit a signal fire and watched the black smoke from oil-soaked rags rise into the blustery September air. They hoped that members of the *Frithjof* or *Express* would see it, though they were not too concerned, as they had nearly a year's worth of food and provisions.

The next day, after a hard night sleeping beside the deflated $100,000 dirigible, the storm calmed, and they could see the *Express* anchored at the foot of the glacier. In time, members of the *Express* and *Frithjof*— including some of Wellman's ground crew—managed to pick their way up the moraine and arrived at the crash site.

It took three days, using ropes, pulleys, and dogsleds, to dismantle and transport the *America* down the precipitous glacier to the ships below. The steel car had suffered some bent tubes and broken wire but was otherwise undamaged. Besides the large tear from the ripping knife, which could be repaired, the envelope also survived intact.

Wellman, Vaniman, and Riesenberg had not had time, during the hectic, frequently terrifying flight and the subsequent salvage operations, to contemplate what they had accomplished. But back at Camp Wellman, it began to sink in. Their flight, though ending ingloriously in an emergency crash landing on a glacier, had been an historic first. Wellman was bitter that the weather had turned on him, but the storm raged on and off for nearly a week, and he realized that even had they cleared the mountains and made it out onto the polar ice, they would likely have been blown down anyway. In those conditions, he admitted, "We could never have made a long voyage . . . We were pretty lucky to get out of it as well as we did."

Wellman knew that some of the press and public, particularly those who had doubted him and his idea from the start, would eviscerate him. And he was right. One Parisian sporting journal, *La Vie au Grand Air,* claimed that his flight had been nothing more than "An American bluff, designed for personal profit and carried out under the cover of impossible scientific conditions." It was bad enough to be excoriated in the aeronautical capital of Paris, but it really stung Wellman when the piece, with its baseless allegations, was later reprinted in *The Seattle Times* and *The Washington Post* in early 1908. Wellman was livid; hadn't he, Vaniman, and Riesenberg just risked their lives in the name of exploration and science? Yes, he was bound to profit from articles about his voyages, but that was the game. Virtually every explorer who survived did so, in articles and lectures. What rankled him most was that his main critic—the one quoted in the Parisian journal, was Alexandre Liwentaal, designer of the original hangar at Virgo Harbor on Danes Island. Wellman had fired him after the summer of 1906, and now Liwentaal appeared to be acting out of vengeance. Wellman immediately had his lawyer file a libel suit against the French journal for $100,000.

To circumvent—or at least provide balance to what he knew would be an onslaught of negative coverage—Wellman sent a dispatch attempting to put a positive spin on their flight and the performance of the *America,* which had in fact, at least for a few hours, flown above the Arctic. His article was picked up by *The New York Times* on September 15, 1907, less than a week after he and his crew had hauled everything down from the crash site on Svitjod Glacier. He explained (perhaps as much to his backers as his readers) that it was mostly foul weather and the faulty compass that prevented them from flying even farther north, but was convinced that the *America* was suited to the task:

> The *America* [proved] her power and capability of being steered . . . The ascent was successful in every respect. The *America* is from every standpoint the strongest airship and the most durable for a long journey that has ever been built. She held the gas splendidly.

After this successful attempt we were all convinced that the *America,* in normal summer weather, can make her way to the pole. We all regard this plan as rational, practicable, and feasible. The thing can be done, and what can be done shall be done.

Wellman never wavered. He instructed his crew, under Vaniman's guidance, to batten down the hangar, camp house, and outbuildings for another winter; two men—including once again the faithful and daunt-less Norwegian Paul Bjørvig—would remain to guard the camp and the valuable tools and equipment. Wellman planned to return immediately to France to begin preparations for another attempt in 1908. What those in the press were mocking and calling a failure he looked upon as "a short trial trip in the snowstorm," one that produced valuable lessons he could build upon. "The season being at end," he wrote, "and winter setting in, the *America* was returned to Paris . . . to be overhauled and improved for the next campaign. We had no idea of giving up the fight."

But Wellman's plan to remain in Europe was soon circumvented by his editor Frank B. Noyes at the *Chicago Record-Herald,* with whom he remained under contract. Wellman, still one of the most intuitive, perceptive, and connected political correspondents working, was recalled from his aerial adventures to cover the upcoming 1908 U.S. presidential election.

As a result of these journalistic responsibilities, Wellman lost some members of his 1907 expedition. Most significant among them was young Felix Riesenberg, who had sought out Wellman because the idea of fly-ing to the pole had been "crazy enough to seem workable," Riesenberg now had second thoughts. During his return to New York, he had plenty of time to consider the last dozen years of his life—twelve at sea as a merchant marine, and the last two in the Arctic with Wellman. He had nearly had his head blown off by his own misfiring rifle; he'd thought he might drown in the polar waters in the overloaded *Frithjof* heading from Tromsø to Danes Island; and the so-called trial flight of the *America* was—whatever Wellman wished to call it—another near-death experience.

When he had first arrived in the Arctic, like many others, he had been consumed by "polar fever," and as he put it, his attitude had been "To the North Pole, or death."

Now, he had spent days and weeks processing the events. In October 1907, he learned that the *Frithjof*, helmed by the jovial Captain Nordby, had collided with a reef and sunk off the coast of Iceland, with the loss of all crew except the chief engineer, who had managed to grab onto a slat of planking and swim ashore. Riesenberg, still just twenty-seven, realized that he might well have been one of those who drowned, and in fact, he was initially listed as one of the dead in early reports of the incident. Seeing his name in print like that and reflecting on his many close encounters with "an Arctic ending" was all too much. When he finally made it back home, his uncle made him an offer: If he could get accepted to university, he would pay for his education. So, Felix Riesenberg applied to the Columbia School of Engineering and Applied Science, and was admitted, turning his varied talents toward civil engineering.[1] Wellman tried to convince Riesenberg to stay on, but Felix, though tempted, had committed to his studies and his promise to his uncle.

Wellman would miss Felix's multitudinous skills, his adventurous spirit and positive attitude, but he understood that Riesenberg needed to move on—he could not just wait around for an entire year to see whether Wellman could scrape together enough backing for another attempt.

Melvin Vaniman, at least, remained totally committed. He had found the entire experience exhilarating. Airship design and flight seemed, in every way, to be his true calling, befitting his penchant for adventure, mechanical design, and aeronautical experimentation. He told Wellman

---

[1] Felix Riesenberg graduated from the Columbia School of Engineering and Applied Science in 1913, and went on to a productive career, becoming chief officer of the United States Shipping Board, superintendent of the New York Nautical School (now Maritime College), and becoming a prolific writer, penning the textbook *Standard Seamanship for the Merchant Service* that enjoyed widespread use. He wrote numerous historical maritime works and novels, and a memoir, *Living Again* (1937), which in part chronicled his experiences with the Wellman *Chicago Record-Herald* Polar Expedition.

that he would remain in Paris and, per their discussions, begin alterations and improvements on *America*'s nacelle, as well as its motor and propellers.

Bolstered by Van's enthusiasm and industriousness, Wellman spent most of 1908 between Chicago, Washington, and New York, simultaneously writing election coverage (correctly predicting that William Taft's Democratic opponent would be William Jennings Bryan) and trying to sustain support for his polar dream. He had several difficult but honest meetings with Victor Lawson, owner of the *Chicago Record-Herald*. Lawson had already put hundreds of thousands of dollars into Wellman's polar expeditions and, despite Wellman's impassioned entreaties, decided against committing further financial support. As a concession, Lawson agreed to loan the airship and all the equipment at Danes Island—most of which he had paid for—to Wellman, for his use should he manage to obtain private funding. Lawson remained the president and controlling stockholder of the *Chicago Record-Herald* Polar Expedition Company, and as such had the power to do whatever he wished with the equipment. For now, leaving it all in Europe seemed a lot more convenient and less expensive than shipping it back to the United States. And who knew—perhaps Wellman might yet succeed, and Lawson's paper could still benefit from the stories he would generate. Lawson was too shrewd a businessman not to keep that potential profit stream open, even as he withdrew any direct fiscal support.

So, Wellman did what he always did—he hustled and promoted his vision. He wrote a booklet titled *To the Pole by Airship*, which outlined his ongoing polar plans and highlighted the endorsement of the French Academy of Sciences and the National Geographic Society. He wrote that recent successes by Count Zeppelin in Germany (a clear appeal to competitive patriotic leanings) and the Wright Brothers at home proved that "mechanical flight machines have roused the whole world to an intense interest in everything connected with navigation of the air." In this he was absolutely right. By now Count Zeppelin's 450-foot *LZ4* had made a twelve-hour nonstop, round-trip flight of 240 miles over Switzerland. Wellman focused his booklet on the airship's possibilities and future, drawing on famous inventor Thomas A. Edison's support of the idea, saying, "Edison has declared that . . . the ultimate solution of the problem is near, and that then the reaching of the North Pole . . . will be very simple."

Wellman tempered Edison's claim by conceding, "The task is not an easy one. I never thought it was. Attainment of the North Pole by any method is extremely difficult. The best proof of this is that the brave and resourceful men of many nations have been trying it for two centuries, without success." But he went on to say, "My faith that this thing can be done is stronger now than it ever was before." Wellman even boldly predicted that not only would his airship reach the North Pole, but "Within a few years . . . the *America* can cross the Atlantic Ocean without much risk of accident with a little help from the winds."

On a new, fancy letterhead—which included a photograph of the airship *America* hovering above the hangar at Danes Island on its maiden voyage and emblazoned with the addresses of his expedition headquarters in Paris and Norway—he sent appeals to major financiers, millionaire aristocrats, and industrial magnates he knew personally. He also orchestrated several behind-the-scenes meetings that were not made public, and by the early spring of 1909 his persuasive efforts were rewarded. A wealthy Russian aristocrat and adventurous balloonist named Nicholas Popov pledged to finance a large portion of the expedition, with the stipulation that he be taken onboard as one of the crew. Two significant silent backers were J. P. Morgan and Andrew Carnegie. Wellman contributed some of his own savings and hired his trusted brother Arthur Wellman as expedition supervisor. Arthur would be charged with heading immediately to Danes Island to ready everything for Wellman's eventual arrival.

For Wellman, it had been a breakneck year, his election coverage and Arctic obsession leaving little time for family and friends, and there were rumors that he had been seeing a young Norwegian woman. Now, once again running short on time, he turned his attention to the last remaining details of the new Wellman Polar Expedition (having removed the *Chicago Record-Herald* from the name). He had to hurry to France to check in on Vaniman's rebuild of the *America,* but first, he needed someone to replace Felix Riesenberg, and a ship to replace the sunken *Frithjof.*

Adding to these pressures, Walter Wellman was by now aware that Robert Peary had managed to raise enough money for yet another traditional dogsled attempt, and he had embarked once more for Greenland in the magnificent *Roosevelt*—repaired and reinforced for ice. But there was also a new, unexpected wrinkle in the race for the North Pole: American explorer Dr. Frederick Cook—who had recently shocked the adventure world by claiming to have been the first to summit Alaska's Mount McKinley (aka Denali), was also in Greenland and announced his own bid for 90° North.

It was getting crowded up north, and Wellman needed to hurry.

9

AIRSHIP OVER
THE POLAR SEA

WALTER'S BROTHER ARTHUR WELLMAN STOOD ON THE DECK OF THE leased steamer *Arctic,* staring dumbfounded as they weighed anchor at Virgo Bay on June 18, 1909. The weather was cold, calm, and clear. That there was little activity was no surprise; only two men, Paul Bjørvig and Knut Johnsen, a seasoned old mariner, had remained behind for the year to care for Camp Wellman. Smoke rose from the chimney of the Wellman House, which was a good sign. Someone was there. But Arthur Wellman could not take his eyes off the "tangled mass of broken timbers, buried under mountains of snow," where the hangar should have been, and he noticed that the American flag on the pole next to the house was hoisted to half-mast.

Arthur went ashore and approached the camp house as men began un-loading the *Arctic.* Dogs barked and snarled warily from their kennel. Just outside the house, Arthur met Paul Bjørvig, who stood with an expression of surprise and relief, his reddish beard long and scraggly, his woolen hat cocked sideways as if he had put it on in a hurry. Arthur introduced himself, and the men shook hands. Bjørvig scuffed his feet in the dirt, kicking rocks with a boot toe, his eyes darting from the hangar to the bay as he spoke.

Bjørvig, speaking in halting sentences, told Arthur that on Christmas Day the previous year, a storm had descended on Danes Island, more violent than any he had experienced in his many Arctic overwinterings.

"If anything can be called a hurricane up here," he said, this had been one. On December 26, though it was minus 15 degrees, he and Knut Johnsen sat shivering. "We did not dare have a fire in the stove," he told Arthur, "because we feared the house would blow down." Gale force winds hurtled in from the southwest, and they sat freezing, listening to the buckling timbers and the whining and screeching of cables as the hangar was blown down, the nine arches toppling one after another, landing with the concussive sound of felled logs thundering to the ground in a forest. There was nothing they could do but wait out the storm and hope that none of the debris from the hangar collided with the house. The storm raged on for six days, then fell dead calm. When they finally went outside, the hangar was destroyed, flattened where it had stood. Its entire roof had been blown one hundred yards away and lay in a crumple of canvas fabric and splintered wood. Fortunately, most of the other buildings, including the house, the machine shop, and the hydrogen-producing apparatus, had escaped the gale unharmed.

Arthur cast his eyes down, taking it all in. Then he looked around, wondering where Johnsen was. He didn't have to ask. Johnsen had died the month before, Björvig told him. He had fallen through the ice while they were out hunting seals. Arthur did not press him for details, and the two men turned from the house to the beach to go help the others unload the *Arctic,* which would return immediately to Tromsø for Walter Wellman, Vaniman, and the modified *America.*

When Walter Wellman saw the *Arctic* arriving at port in Tromsø a week later, with its flag flying at half mast, he knew something was wrong. He walked to the end of the dock and saw, to his surprise, Paul Bjørvig disembarking the ship. He had expected Bjørvig to remain with Arthur on Danes Island, to help ready everything for the final shipment. Then Bjørvig told Wellman what had happened. He and Johnsen had been out on skis, hunting seals on grounded ice beyond the headland near camp. When the ice became too slick they had taken off their skis and carried them. Johnsen had been just a few yards ahead of Bjørvig when Johnsen slipped down a steep decline and into the water. "When I looked down I saw Johnsen lying between two huge pieces of ice," Bjørvig said, his eyes wet with tears as he remembered the scene. "He was lying on his back and scrambling with his hands . . . I heard him say, Paul, help me!"

When Bjørvig had looked down below, Johnsen was floundering and gasping in the icy water thirty feet below, the sheer ice as smooth as a wall. Bjørvig tried to throw him a rope, but Johnsen fell beneath the surface. "The skis drifted slowly out to sea," Bjørvig said, "I sat myself down and looked at the place where he had disappeared, and thought of the sorrowful fate that had struck this . . . elderly man."

Wellman was quiet for a long time. He patted Bjørvig on his shoulder, thinking of all the man had been through, all he had done and sacrificed for him. How he had lain beside the frozen body of Bernt Bentsen during the dark, cold winter of 1898–99 on Franz Josef Land. When Bjørvig finally gained his composure, Wellman asked him of his plans, and whether he wished to return to Danes Island with him aboard the *Arctic,* which would be leaving as soon as they'd loaded it.

Bjørvig surveyed the bustling island city, the waters in the port shimmering in the June sun. This was his home. While wintering at Danes Island for Wellman in 1906–07 with Felix Riesenberg, he had received word informing him that one of his elder sons had died on the ice while out bear hunting. But he still had a wife, two sons, and three daughters here. He would go to them. Wellman thanked him for his courage and sacrifice, paid him his wages, and the men embraced, then parted ways.

"I have had enough sorrow from the Arctic," Bjørvig reflected, adding, "But if a man has no sorrows, he has no joys."[1]

Walter Wellman and Melvin Vaniman arrived at Danes Island three weeks later on the "slow and uncertain" three-masted schooner *Arctic.* Wellman

---

1 Born in 1857, Paul Bjørvig lived until 1932. He died at his home in Tromsø. He was buried in the cemetery there, where his headstone is commemorated with a polar bear representing his life in the Arctic. Above his name is the word "Ishavsfarer," meaning "seafarer."

had hired and sent ahead forty-five Norwegian carpenters from Tromsø, offering to pay them all bonuses if they could repair and erect the destroyed hangar in a month. In the end, it took them nearly six weeks, but Wellman paid them their bonuses anyway.

Vaniman had made several alterations and improvements to the *America*. He installed an additional eight-cylinder 125 horsepower E.N.V. Antoinette engine, which augmented the Lorraine-Dietrich (80 horsepower) to two engines and driving systems and affixed rotating booms that could redirect propeller wash in multiple directions. He also implemented a better and more efficient "triple box rudder system," swapping it out for the single rudder used in 1907. Wellman fully believed his revamped *America,* better and stronger than ever, could now push between twenty and twenty-five miles per hour and was worthy enough "for another aerial onslaught upon the Pole. . . . To voyage across the great white place on the Arctic charts, the blank marked 'unknown.'"

Wellman would be captain and navigator of the 1909 attempt. Vaniman would again be chief engineer. The new crew included Nicholas Popov, the wealthy Russian aeronaut whose generous financing had made this attempt possible. Popov was quirky and fastidious, a man rumored to have had his laundry sent by ship to London because he believed their services superior to those in St. Petersburg. At Melvin Vaniman's urging, Wellman had hired Vaniman's "cool headed and resourceful" brother-in-law Louis Loud, and finally A.J. Corbitt, a highly skilled British airship specialist who had joined Vaniman in Paris.

By the first week of August, the hangar repair was complete, and everything appeared ready. Under Arthur Wellman's management, and with Vaniman's experience and expertise, Camp Wellman had once again become a "scientific village in the Arctics," as a perceptive London-based reporter had dubbed it. Once more, the *America* would carry sleds and sled dogs, a small boat, tents, a year's worth of provisions, everything required for an extended stay on the ice should calamity strike. Wellman reasoned that should they fall just short and get within the vicinity of the pole, they could convert into a traditional sledging party to go the rest of the way to claim the pole, "having only the return journey to make instead of the upwards and back trips attempted by many of us in the old ways." If necessary, they could use fabric from the nacelle covering, or the envelope, for shelter on the ice to overwinter, returning the following spring.

On the morning of August 15, after days of anxious watching and waiting, the winds were finally deemed favorable for a launch, with a light breeze coming from the south. Wellman convened the crew and told them that A.J. Corbitt would remain at base camp to assist, not wanting any extra weight and another person moving around in the airship car. Corbitt was disappointed, but he agreed to aid in any ways needed, and Wellman asked that he oversee the towing of the airship out into the bay by small motorboat.

At 10:00 A.M., Wellman called for the airship ground crew to prepare for departure, and he, Vaniman, Nicholas Popov, and Louis Loud took their places in the *America*: Wellman and Popov up in the wheelhouse, Vaniman and Loud at the engine room.

Once more, a few dozen men guided *America* slowly out of the hangar, clinging to hold-down ropes and awaiting instructions. When they had just cleared the hangar a strong swirling gust drove the airship back, and it nearly collided with the hangar entrance. They cleared the top by a few feet, and Vaniman started one of the engines. It was a brilliant, clear morning with no ominous clouds, and tethered to Corbitt's motorboat, *America* moved slowly out over the water of Virgo Bay.

Arthur Wellman ran along the beach shooting motion picture footage as *America*'s engines propelled the airship faster, riding sixty or seventy feet above the water, and soon Corbitt was the one being towed, skittering along behind, his boat making a larger and larger wake until he yelled for them to cut loose. The *America* lifted then, free and under its own power, soaring into rising and swirling wind drafts. Wellman navigated through the east strait toward Amsterdam Island's Smeerenburg Point, where a sudden wind blast drove them just above the rocks. But Wellman calmly brought the airship around. The new engines, propellers, and rudder responded well, and the trailing equilibrator skipped along behind, leaving a line in the sand near the Dutch whalers' graves "where the summer town reeked with whale oil and rum in the long ago." Then they cleared the point and were safe above open water once more.

Looking back toward camp, Wellman and his crew could see men on the beach and on the hills of Danes Island waving their hats and cheering. The airship was cruising smoothly, well balanced, experiencing some engine vibration but minimal pitching, yawing, or rolling, and after fifteen to twenty minutes, Wellman's dirigible was making its way north, gaining an altitude of nearly three-hundred feet. He could see the *Arctic* following along behind, trying to keep pace, but soon it appeared as just a speck on the polar sea, the dark waters dotted with small islands of white pancake ice. Wellman grinned and laughed, nudging Popov who stood next to him, taking it all in. To the east, snow-clad mountains and blue-white glaciers gleamed in brilliant late-summer sun. Beyond, to the north, the immeasurable ice pack was coming into view, the entire scene an exquisite panorama of earth and water, ice and sky.

Wellman gave Popov the helm and checked the instruments. They were making just over twenty-five miles an hour, aided by a steady southerly wind. He scribbled some notes and calculations, marked the time on his chronometer, then looked north and smiled. At this rate, they would reach the pole in under thirty hours. Elated, Wellman climbed his way to the upper deck and "hung there suspended between the heavens and the earth," his mind reeling and dreamy with possibility. Then he returned to his many tasks. His hand-held standard compass read steady and accurate. He went back to the wheel, where Popov was beaming at the helm. "The slightest movement of the wheel," he said, "was sufficient to turn her up the wind." Wellman nodded, noting that the mounted steering compass seemed slightly erratic, but he attributed this not to magnetic anomaly or failure, but to the ship's engine vibration.

Vaniman came up from the engine room. "What's the course?" he yelled above the roar of the engines and the whoosh of the propellers.

"True north," Popov yelled in reply.

After nearly two hours of smooth flying, the northern coast of Spits-bergen was receding from view, forty or fifty miles behind them. Above the horizon, wispy tendrils of cirrus clouds fanned out like wing feath-ers against the cerulean sky. Ahead was a seemingly endless unfurling of white polar pack, flecked with amassing ices floes, spanning all the way across the polar sea, to the North Pole and beyond. At that moment, Well-man looked out the window, saw a blur of something falling away from the *America,* and cried out, "My God, look at that!"

Just as he realized it was the twelve-hundred-pound equilibrator plung-ing to the sea and ice cakes below, he felt a jolt, and the *America* shot up-ward, rising at terrifying speed into the clouds.

10

LAST GASP OF THE POLAR DREAM

LOUIS LOUD PANICKED AS THE *AMERICA* HURTLED UPWARD. HE pushed past Vaniman and grabbed the cable leading to the hydrogen-gas release valve. Despite his brother-in-law's protestations, Loud refused to relinquish his grip, but Vaniman convinced him not to pull on it, not yet. He needed to talk to Wellman. All men aboard could feel the air temperature drop as they ascended fast, gaining five thousand feet of altitude in minutes. Their ears popped and rang with the fast elevation gain, and the higher they went, the harder the winds blew.

Wellman left Popov at the wheel and scrambled aft to consult with Vaniman, who had finally managed to wrest the release-valve cable from Loud. They had to act fast—too much altitude too quickly could cause the envelope to explode. With Wellman's consent, Vaniman pulled the valve-line and released enough hydrogen gas to slow their ascent. As they leveled off, Wellman ordered Vaniman and Loud to lower the retardateur, hoping its weight would replace that of the equilibrator, which they deduced had sheared off somewhere near where it attached to the airship car. The two men struggled for some time but managed to guide the four-hundred-pound steel snake downward, and with that and the released hydrogen gas, they started descending.

Wellman now had a crucial decision to make. At this higher altitude, strong winds were pushing them northward even faster, and soon they were some seventy miles beyond the northern Spitsbergen coast, heading due north, exactly where they wanted to go. But with the loss of the equilibrator, and the released hydrogen (already the somewhat deflated envelope was losing its shape), Wellman was uncertain how he could properly control his airship. And the half-ton equilibrator had contained most of their reserve food, which they would need should the airship go down hundreds of miles out onto the polar ice.

Wellman was at an impasse. At this rate, should no further accidents occur, they might conceivably be at the North Pole within the next twenty-four hours. It was what he had come here to do. Wellman looked behind them and saw "the whole northern part of Spitsbergen spread out in one great frozen picture . . ." receding in the distance. Before him was the great unknown. In the freezing air, the engines still droning along beautifully, he silently wrestled with his choices: hold their course steady north, and risk all their lives should they crash or alight on the ice and run out of food; or turn around and try to make it back.

Reluctantly, Wellman made his way forward to the wheel and told Popov, "Turn her around and steer for Spitsbergen." Popov complied, struggling to maneuver as strong winds pummeled the dirigible. Once heading southwest, Wellman took the wheel and instructed Popov to start moving gear and the remaining food stores aft, as close to the rear as possible, for better weight distribution and balance. This helped, and with Vaniman valving off more hydrogen gas, they began descending and found calmer air, though they now fought a strong headwind, their speed reduced to just eight to ten miles per hour.

And then, inexplicably, the engines quit, and they were driven back north for a time, with Wellman fighting to keep *America*'s nose pointed directly into the wind. During this time, drifting powerless at the mercy of the gales, Wellman no doubt thought of S.A. Andrée, who had also left from Danes Island and was last seen drifting in the same direction. His body—and those of his two comrades—were likely frozen somewhere in the vastness of ice below.[1] Fortunately, Vaniman managed to start the motors again, and he and Loud then pumped the internal ballonets full to help maintain the envelope's shape, and soon they were heading southwest once more, the airship responding well at the helm, descending as they motored along, with Spitsbergen's northern coastline coming back into view.

Scanning ahead with field glasses, Wellman saw a ship heading their way, weaving through leads between ice floes. He altered course slightly to steer toward it. By now, the retardateur, designed to arrest them as a kind of ice anchor and which they had employed as a makeshift equilibrator, was contacting the ice below. But the retardateur had a long loop of steel cable at its end, and that loop began snagging on jagged ridges of ice floes, halting their progress. Wellman and Popov worked together, hour after hour, swinging the ship in complete circles to free the steel loop from the ice, until they could continue flying forward. It was tedious going, but they made headway until after a few hours, they could clearly see a ship halted at the edge of the pack ice, which it apparently could not penetrate.

It was not Wellman's hired ship *Arctic*; he did not recognize the vessel. But it was close enough that though Wellman believed he could have guided *America* back to Danes Island, for safety and expedience, he steered directly toward the steamer and told Vaniman to valve off more gas to lower them closer. When near enough, Wellman could see men waving from the deck of the ship, and he ordered Vaniman to drop a towline down to them.

---

1 In fact, the balloonists were indeed frozen. Wellman was right about the fate of *Andrée* and his companions, but it would be more than thirty years before the full story was revealed. In August of 1930, a crew on a Norwegian scientific expedition, aboard *Brattvag*, discovered the camp and the remains of three men on White Island (Kvitøya), 40 to 50 miles east of Svalbard's North East Land, which they had reached on foot after their balloon went down. Their diaries were discovered, detailing their ordeal on the ice. Rolls of exposed film were also discovered and were able to be developed despite decades buried under ice and snow. The developed photographs, and their diaries, allowed most of their journey to be reconstructed, though the exact cause of their deaths remains mysterious. Andrée, Strindberg, and Fraenkel were transported to Stockholm and buried there on October 5, 1930. An interesting book about the flight of the *Örnen* (*Eagle*) is *The Ice Balloon* by Alec Wilkinson (Knopf, 2011).

Men aboard the ship secured the towline, and the Norwegian coast survey steamer *Farm* (not to be confused with Fridtjof Nansen's much more famous *Fram*) began towing them slowly back toward Virgo Bay. The winds increased, buffeting the *America* violently side to side, and Wellman worried that the jolting might tear the car from the envelope, or bend the steel framework of the car. He advised Vaniman to bring them down into the water. As the airship car slapped onto the surface, wind-whipped waves washed over the car, and the dogs howled like wolves. The *Farm*'s captain, Gunnar Isaksen, sent lifeboats to help, fearing Wellman's crew would be capsized and drowned. When the lifeboats pulled abreast of the car, Wellman was sitting at the wheel smoking a cigar, directing his own crew to begin offloading gear.

Vaniman put *America*'s lifeboat over the side of the wave-rocked car, and working together with the *Farm*'s lifeboats, after several trips through roiling waves, most of *America*'s gear—scientific instruments, tools and provisions, and the dogs—was moved to the *Farm*. Finally, Vaniman, Loud, and Popov left the car, with Wellman—perhaps symbolically— the last to abandon his airship. Wellman greeted and thanked Captain Isaksen, who told him he and his crew had watched the *America* flying for hours, and when they'd seen it rocket up into the clouds, had followed along in case they needed help.

Within a few hours, Captain Isaksen had transported Wellman, his crew, and his beleaguered dirigible back to Camp Wellman, where they made anchor just offshore. Arthur Wellman and A.J. Corbitt were there to greet them and to assist in bringing everything ashore. While Wellman told his brother Arthur of their adventures, Corbitt hopped into the airship car to help Vaniman disassemble the engines and gas tanks, and disconnect the car from the envelope, still inflated with nearly 250,000 cubic feet of hydrogen. Vaniman drained fuel from the fourteen tanks, first fore and then aft—puncturing each with a pick axe—to lighten the car and make it easier to bring onto the beach. As fuel drained from the tanks, a wind-burst caught *America,* and Vaniman quickly hopped off. The nose of the airship lifted high, rising in the wind as men yelled for Corbitt to jump. When he finally leapt, he fell twenty-five feet, landing in sandy, loose stones that broke his fall enough to prevent serious injury.

As Corbitt stood brushing himself off, the envelope tore free of the air-ship car and sailed like a giant distended kite into the sky. Everyone watched in disbelief as it soared skyward, up and up until it was more than a mile above them. "Its hydrogen expanding in the thinner air at that altitude," recalled Corbitt, it "burst with the roar like a cannon" and began falling back to earth. Torn pieces of fabric landed on the *Farm,* and fearing the rest of the six-thousand-pound rubberized envelope would come crashing down on them, some sailors leapt overboard into the icy water. Luckily, the envelope missed the steamer by only one hundred feet, splashing down mostly intact. "It was the most thrilling thing I'd ever seen," said Corbitt.

Wellman was less thrilled. The envelope, while mostly salvageable, would need extensive repairs, either at Danes Island or in France. The na-celle had been damaged by its twenty-five-foot fall onto the rocky shore, its steel tubing bent and broken. After inspecting it, he deemed it unworthy, and ordered Vaniman to strip it down and retain any useable parts. After thanking Captain Isaksen for his assistance and saying goodbye, Wellman ordered his crew to pack everything into storage in the camp's outbuild-ings. Then, he and Arthur went to the Wellman House to discuss every-thing that had happened and to contemplate the future.

Walter Wellman was overwhelmed by "bitterness inexpressible" at his most recent setback. "Would the Arctics never bring me anything but bad luck," he wondered. He could take some solace in knowing that, round-trip, he had flown a motor-driven dirigible some 120 miles over the polar sea, an historic record. He also understood that all such pioneering ven-tures were a process, each attempt and failure providing experience and building on the next. How many times had Peary tried for the pole? For more than a decade, at least three, Wellman figured, counting the current expedition now in progress. Peary had lost eight toes to frostbite on one attempt, and still he kept going. Wellman also had no intention of giving up. He also knew he had to preempt the doubts and criticism he would face when word of his recent failed enterprise reached the world, which would be immediate.

"I do not give up the fight," he wrote in a dispatch to *The New York Times* on August 22, 1909. "Had the ship not broken up while landing," he added with bravado, "we would have tried another voyage this month with a new guide rope." That may well have been true. The cruel irony of the fickle Arctic weather added salt to his emotional wounds; for four consecutive days after returning to base, southerly winds had blown consistently to the north, and had they kept going, they might well have made it to the North Pole—or quite near it—within a day of his decision to turn back. But Wellman rationalized his decision by saying he could not in good conscience have endangered his crew in a wounded dirigible. "My own life, yes," he wrote. "Theirs, no."

But this year's attempt was over, and Wellman prepared to depart Danes Island once more. Before leaving, he instructed the carpenters to begin expanding the hangar, "determined to continue the quest for the Pole with a new and enlarged *America* . . . with which we intend going . . . in the early summer of 1910." As for Wellman's future crew, adventuresome Nicholas Popov found the entire experience exhilarating, but said he was ready to try something new; now he wanted to go learn to fly one of the Wright "aeroplanes" he had been hearing so much about. Melvin Vaniman remained entirely committed, as did his brother-in-law Louis Loud and A.J. Corbitt.

As he waited for the steamship *Arctic* to retrieve him from Spitsbergen to return to mainland Norway, Wellman contemplated the coming year, and worried privately about word of Robert Peary and Frederick Cook, either one of whom might soon be returning from their brutal slogs behind teams of dogs across the buckling, deadly, labyrinthine ice. More than ever, he believed that "the race for the Pole was a struggle between the old and the new methods—the sledge versus the airship." Wellman remained confident that the airship would prevail.

The *Arctic* arrived on schedule near the end of August, and Wellman and his crew left immediately for Tromsø, where they transferred to a faster coastal steamer. One bright and crisp early September morning, while tied up in the western coastal town of Bodo, Wellman was up on the deck chatting with A.J. Corbitt, taking in the sunshine, the bustling port, the mountains all around them. As they stood at the rail, Wellman smoking a cigar, a messenger walked up and handed Wellman a telegram. Wellman took off his wire-rimmed spectacles, cleaned them with his handkerchief, put them back on and read the telegram. Wellman reached into his pocket, found some coins, and tipped the messenger. "No reply," he said gruffly.

"Not very good news, sir?" Corbitt asked, reading the expression on Wellman's face.

"No," Wellman replied, exhaling in a sigh of resignation. "We shall not be going to Spitsbergen again next year. Dr. Frederick A. Cook has found the North Pole."

11

CONTROVERSY AND A NEW DREAM

WHEN WELLMAN REACHED COPENHAGEN EN ROUTE TO PARIS, HE found the Danish city "ablaze with flags and flowers in honor of Dr. Cook, who had arrived the day before." The city, and indeed much of the world, was celebrating Cook's grand achievement, a dream sought for centuries, a quest, as Wellman thought of it, to satisfy man's "all-compelling instinct to know all of the unknown." But on his sleeping car train journey from Oslo (then called Christiania) to Copenhagen, Wellman had been restless, his mind uneasy about the news of Cook's conquest.

Dr. Frederick A. Cook had bona fide credentials as an explorer. The American medical doctor had participated as surgeon and photographer during Robert Peary's 1891 expedition to North Greenland, had led two of his own Greenland expeditions in 1893 and 1894, and, significantly, had participated—along with Norway's Roald Amundsen—in the first overwintering in Antarctica aboard the *Belgica* in 1898. More recently, in 1906 he claimed to have been the first to summit Mt. McKinley (Denali), the highest peak in North America at 20,310 feet. But numerous people— experienced Alaskan mountaineers among them—now seriously doubted Cook's McKinley summit claim. The lone photograph he took wasn't con- clusive. "It could easily have been taken almost anywhere above the snow- line," argued a noted a glaciologist and professor of geology from Cornell University. The photograph was being widely dismissed as a fake. Most damning of all, Ed Barrill, Cook's sole climbing partner, admitted to close friends that the two men ascended to no higher than five thousand feet.

As a result of these doubts, in Copenhagen, Wellman—who during his layover was within a minute's walk from Cook's hotel—chose not to go congratulate him. By now Wellman had read Cook's complete account, as it was first cabled and published in *The New York Herald* on September 2, and based on what he read, "Partly through intuition, and in part through logic, I felt sure his tale was not true." At the moment, Wellman chose not to provide details for why he did not believe Cook, but something about Cook's character made him suspicious. Wellman later elaborated, "If I had had faith in him, it would have been my pleasure as an American citizen to offer him my congratulations upon his great achievement; and as an Arctic man it would have been my duty to do so. But under the circumstances I could not shake his hand, and proceeded to Paris without seeing him."

En route to Paris, Wellman began rethinking what he had told A. J. Corbitt—maybe they *would* return to attempt the North Pole by airship once more. It would depend on whether Cook could provide indisputable proof, which Wellman highly doubted. If Cook failed to incontrovertibly make his case by providing diaries, logbooks, and charts, perhaps the prize was still there for him to seize.

In Paris on September 6, while contemplating retrieving the burst *America*'s envelope from Danes Island for repairs, Wellman was handed a copy of *The Evening Star*. In bold ink splashed across the front page was the headline: "Peary Sends in Report that American Flag Is Nailed to the North Pole." Wellman scanned the article, discovering that Peary had cabled a short dispatch from Newfoundland, stating bluntly:

"Stars and Stripes nailed to North Pole. PEARY."

For Walter Wellman, at that moment, his polar dream had officially ended. His gut told him that Peary had done it. He knew the man, knew he had devoted more than two decades of his life to that singular obsession. But even more than that, as a newsman and a polar explorer himself, Wellman knew what would happen next. With two separate claims coming within days of one another, there would be endless arguments, claims, and counterclaims. The sponsoring newspapers were deeply entrenched, and things were going to get ugly. "There will be so much controversy over this," he said, "That there will be no more honor in even trying again."

He could not have been more prescient. Arriving back in the United States, Wellman pored over newspaper after newspaper from every corner of the publishing world. The Cook–Peary Controversy was spinning out of control. Cook and Peary started mudslinging at each other in a public feud, each questioning the other's methods, route, and character. Cook cleverly took a very public victory lap, embarking on a lucrative lecture tour. Nearly everyone had an opinion, and everyone took a side, from Danish King Frederick VIII (for Cook) to President Theodore Roosevelt (for Peary). What might have been a cool-headed evaluation of each man's logs and records (which were suspiciously slow in coming from both men) quickly became a news frenzy, a bout between *The New York Herald* in Cook's corner and *The New York Times* in Peary's. The controversy raged all that autumn and became a sensational brawl that devolved into a national embarrassment, both men being Americans.

Wellman himself got into the fray. As one of the three men who had been vying for the pole at the same time, he was hardly a dispassionate or uninformed observer. The claims by Cook and Peary had squelched his plans for another bid, so he had some emotional investment in discovering which of them, or whether either, had beaten him to the pole. In a lengthy and comprehensive *New York Times* article written later that autumn, Wellman looked carefully at the narratives and records each man had so far presented. Peary had by now submitted his travel logs and instruments to the National Geographic Society, which had determined his claims valid. Cook was now saying that the bulk of his documenting materials and instruments had been left in Greenland, and that he would soon send for them.

Based on what Wellman had to go on, he opened with a damning headline: "Wellman Riddles Cook's Narrative." Wellman found Cook's stated daily distances impossible given his starting point at Annoatok, northwest Greenland, covering seven hundred miles in sixty-two and a half days: "No such sledging feat had ever been done, nor anything approaching it." Wellman pointed out that Cook was "vague and indefinite," had neither the man or dog power (three men, two sledges twenty-six dogs versus Peary's twenty men, nineteen sledges, and a hundred thirty-three dogs) to make the distances he claimed. Wellman concluded that the only explanation was that Cook had fabricated his journey and, worst of all, had likely written it beforehand.[1]

Wellman's assessment was soon corroborated. Near the end of Cook's lecture tour, in early December 1909, he finally submitted records to the University of Copenhagen for expert evaluation, then went into hiding, with close associates saying that he had suffered a nervous breakdown. An explosive *New York Times* article followed shortly on December 9, publishing confidential materials they had obtained that suggested he had hired people to falsify his records. "Fraudulent Observations Made for Dr. Cook Before His Records went to Copenhagen; Sworn Testimony of the Men Who Made Them" read the headline.

Less than two weeks later, the Danish committee brought forth their

---

1 The controversy raged for decades, resulting in diehard "line-in-the-sand" factions on either side and dozens of books on the subject. Today, most dispassionate observers concede that almost certainly neither Cook nor Peary made it to the North Pole, and that unfortunately, both men lied and falsified their records. In 1985, British explorer Wally Herbert, who had himself sledged to the North Pole in 1969 and during his career covered over 20,000 miles retracing the routes of both Peary and Cook, was for the first time granted access to the "Peary Diary," Peary's journal from the 1909 expedition. After exhaustive scrutiny of the diary, journals, and comparisons to subsequent attempts—some even using snowmobiles—Herbert concluded that Peary's claimed speeds and distances were not possible and that he had come up short of the North Pole by thirty to sixty miles. On August 23, 1988—moved by Herbert's findings—*The New York Times* issued a remarkable "correction" to their reporting at the time of the 1909 Peary Expedition. "Most authorities today agree," their editors wrote, "that no one has an untainted claim to be the first—neither Cook, nor Peary . . ." For those wishing to go down the North Pole controversy rabbit hole, see *The Noose of Laurels* by Wally Herbert; *Cook & Peary: The Polar Controversy Resolved* by Robert. M. Bryce; and for a concise round-up, the recent and excellent 2023 book *Battle of Ink and Ice* by Darrell Hartman.

unanimous decision sealing Cook's fate and reputation: "Not Proven."[2] Frederick A. Cook was immediately disgraced, persona non grata in the exploring world. The Arctic Club of America, for which he was currently serving as vice president, barred him from seeking reelection. The Explorers Club immediately and officially discredited his McKinley summit record. Cook receded into exploration anonymity, slinking off to travel secretly through South America and Europe. Meanwhile, Peary was subsequently granted, at a Congressional hearing, a promotion to rank of rear admiral, a full pension, and was officially recognized for having been first to reach the North Pole.

With the contentious Cook–Peary controversy seemingly settled, Walter Wellman pounced on the vacuum left in the press. On February 1, 1910, he announced in *The New York Times* that in July, in just six months, he would attempt something new that had never been done: he would fly an airship across the Atlantic Ocean.

---

2 Ten years later, in 1919, Cook founded the Texas Eagle Oil Company, and afterward the Petroleum Producers' Association; both became embroiled in what one observer called "A giant stock reloading scheme" similar to a Ponzi scheme. In April of 1923, Cook was indicted on numerous counts of fraud and sentenced to fourteen years and nine months in prison. He was sent to Leavenworth Penitentiary in 1925 and released in 1930, only halfway through his sentence, having proven to be a model inmate with exemplary behavior. See Julian Sancton, *Madhouse at the End of the Earth*, 296–307.

12

THE BIGGEST THING
IN THE WORLD

THE PACE AND SCALE OF WALTER WELLMAN'S TRAVELING, JOUR-
nalism, airship endeavors, and "family life" were dizzying, especially considering that in the early twentieth century there were no transoceanic
or transcontinental flights, and the fastest transatlantic ocean crossings
during the Golden Age of the Ocean Liner took five days to a week. The
Mauretania set the crossing record in 1907, making the trip in five days,
five hours, and ten minutes, a record that held for thirty years, until the
Queen Mary bested it by nearly a day. Wellman was constantly traveling aboard such liners, cruising back and forth between Europe and the
United States to conduct his various business and personal affairs.

Soon after announcing that he would attempt to navigate his dirigible *America* across the Atlantic, he returned to France to oversee Melvin
Vaniman's work on the airship's reconstruction. But Wellman had other
obligations, too, not the least of which was spending a little time with his
two-year-old son, Walter Jr., and his newborn daughter, Elsa. As rumors
had suggested, Wellman had entered into a relationship with a Norwegian
woman named Bergljot Bergersen. Few details are known about where
they met or how often they were able to spend time together. What is
known is that Ms. Bergersen was twenty-six when she gave birth to Walter Wellman Jr. in Washington, D.C. in 1908, and their daughter, Elsa,
arrived in France early in 1910.

Just how Wellman Sr. explained any of this to his wife, Laura, is also not known. If he tried to keep the affair a secret, he did a poor job of it, for published pictures exist of Wellman holding toddler Walter Jr. in front of the airship *America* in 1910. At any rate, he wasn't around to see his mistress and children for long in France, because by March of 1910 Wellman was racing up the White Nile in a chartered boat, pursuing a major story for the *Chicago Record-Herald*.

Former President Theodore Roosevelt was known to be somewhere on the river, returning from a safari he had taken after his last term in office ended. Many of the world's most adventurous newsmen had set out into the heart of Africa to meet the enigmatic former president and discover his future plans. Wellman was among them. He admitted he was spread rather thin, as usual. "From an airship voyage over the Arctic Ocean in August 1909 to a steamship race up the White Nile almost to the equator in March 1910 was rather a quick transition," he conceded. But for Wellman, such a harried pace was nothing new.

Wellman had hired the fastest steamer available to "race rival newspaper men several hundred miles above Khartoum for the satisfaction of being the first to meet the ex-President." Wellman bested them all, finding Roosevelt standing barefoot in pajamas on the deck of his boat. His face was darkened by months in the sun, and he appeared trim and healthy.

Roosevelt greeted Wellman like the old friend that he was, and invited him onto his steamer, where they had breakfast. They talked about Roosevelt's experiences hunting rhinoceros, specimens of which he was bringing back to display in museums in the United States. Roosevelt, ever the adventurer, wanted to know details of Wellman's airship flights over the Arctic, and as a naturalist and hunter, he wanted to talk about polar bears and walruses and whales. They spoke easily until Wellman changed the subject to politics, since that was, at least in part, what his editors and the public was interested in. Would Roosevelt seek a third term?[1] "I have

---

1 Term limits for U.S. Presidents would not be established until the passing of the 22nd Amendment to the Constitution on February 27, 1951.

nothing to say and will have nothing to say on American or Foreign politics," Roosevelt said flatly. "No political interviews of any kind." With that,
Roosevelt invited Wellman to remain on his steamer and travel with him
back to Europe, an invitation Wellman readily accepted. Off the record,
Wellman joked, "We sat down and settled all the affairs of all the nations
to our mutual satisfaction."

As they continued downriver, Wellman swelled with swagger and panache. "I had written my cablegram describing the race and my triumph
before the smoke from my rival's boat was seen downriver, puffing toward
us as fast as she could with a party of discomfited journalists aboard her."
Wellman had the coveted scoop, and remained with Roosevelt for weeks,
traveling through Europe and eventually arriving in London. They parted
ways there, and Wellman was finally back in the United States by early
June, 1910.

At least, for an airship Atlantic crossing attempt, Wellman did not have the
short Arctic summer to contend with. This was providential because the
airship wasn't ready yet. While still in France hobnobbing with Roosevelt
and visiting his mistress and their children, Wellman had checked in on
Vaniman and pressed him to ship *America*'s envelope, nacelle, and engines
to the United States by August. There, they would finish any remaining
work at the impressive, newly built wooden hangar near the boardwalk's
end in Atlantic City, which would house the now 228-foot-long *America*.
Vaniman's reconfigured airship would now contain 345,000 cubic feet of
combustible hydrogen gas, an 80 horsepower engine to be used in favorable winds and a 200 horsepower engine for bucking strong headwinds,
with a gross weight, including food, fuel, cargo, and crew, of nearly 30,000
pounds.

Wellman employed his charisma and persuasive powers to engage three
separate newspapers in shared sponsorship of the Atlantic crossing venture: The *Chicago Record-Herald* (Wellman had finally convinced owner
Victor Lawson to provide financial support), *The New York Times*, and
The London Daily Telegraph. Everyone—editors, the adventure-hungry

reading public, and experts in the aeronautical world—knew that this unprecedented proposed three-thousand-mile undertaking was extremely dangerous, and that was at least part of the allure. The reality was, whether Wellman succeeded or failed, whether he and his crew lived or died, this would be one of the most anticipated and thoroughly documented stories of its time.

Wellman knew it too. Though they had taken all precautions necessary by covering the engines with asbestos and placing the motors far from the gasoline tanks, severe risks remained. "Once well on our way," he wrote as the journey neared, "the danger of fire or explosion will be ever present in our minds. The combination of a ton of hydrogen, nearly three tons of gasoline, sparking motors, electric light, and wireless is not one to inspire perfect confidence." Hazards too numerous to list were also possible, but he had thought of a few: lightning strike could ignite the hydrogen; a sheared propeller might pierce the gas envelope; the engines could fail.

Regardless, Wellman, Vaniman, Louis Loud, and three additional crewmembers had all agreed that the potential risks were worth the potential gains: not only making history, but changing the scope of manned flight, creating a paradigm shift from land or sea to long distance aerial transportation.

For Wellman, the potential monetary rewards were staggering. According to Benjamin Hampton, editor of *Hampton's Magazine*, success would reward him handsomely: $75,000 to $100,000 for world rights to the magazine story; up to $50,000 for book rights; and an estimated $100,000 for a lecture tour. "All in all," wrote Hampton, "a successful trip . . . will probably net him $1,000,000 in two or three years. I believe the thing he is trying is the biggest thing in the world, outside of the discovery of the North Pole." And that, at least as far as everyone believed at the time, had already been accomplished.

In the predawn darkness of Saturday October 15, 1910, Walter Wellman sped through damp city streets and arrived by automobile outside the airship hangar in Atlantic City, accompanied by his wife, Laura, and two of their daughters. He hurried into the dimly lit hangar, where Vaniman greeted him. "It's Europe or bust," Vaniman said, shaking Wellman's hand. The fully inflated *America,* improved and innovated almost entirely by Vaniman, rocked and tugged above the ground, straining against the hundreds of sandbags holding it down. Soon they could hear the voices and footsteps of some one hundred Atlantic City fireman and policemen arriving; they'd volunteered to guide *America* out and deliver it to the motor yacht *Olive,* which would tow the airship out into the Atlantic waters for release. Dense fog shrouded the charged scene, but the weather forecast was promising, the best they had seen in weeks, according to the U.S. Weather Bureau: "Fair Sunday and Monday, with moderate westerly winds, becoming variable." But Wellman had read numerous reports, and he knew that a tropical storm was active in the Gulf of Mexico, which gave him pause, despite it being a considerable distance to the south.

Wellman and his airship had been vexed by changing weather before, to disastrous ends in the Arctic. Although Wellman had been assured by the Willis L. Moore, chief of the United States Weather Bureau, that "There was no danger of the airship meeting the West Indian hurricane now coming this way," even hearing the words "tropical storm" and "hurricane" was unnerving. But Moore was insistent, reiterating that "There is no possibility of the *America* falling into the path of the tropical storm, and *America* ought to have first-class weather conditions during the next day or two." It was all Wellman had to go on, and he hoped the Bureau chief was right.

After two hours of final inspections, all six men—Walter Wellman, Melvin Vaniman, Louis Loud, Murray Simon, Fred Aubert, and Jack Irwin—were ready. They shook hands and wished one another good luck. Wellman, by now fifty-two, "of somewhat ample build, with graying hair," gave instructions as sandbags were released and the orderly line of volunteer police and firemen marched the ship down to the waterfront. Even in the damp, dreary weather, a crowd of thousands had gathered along the boardwalk and seashore, come to witness history. Once the ship had cleared the hangar, Wellman strode over and hugged his wife and daughters goodbye. Though he had put them through much over the years, they had remained loyal to him, and were humbled and awed by his ambition and courage in this latest undertaking; all knew it was a real possibility they might never see each other again. Then Wellman turned from them and boarded the dirigible.

"Up and across," the crowd cheered as he waved from a small door amidships.

Vaniman trotted some distance away, far enough to view the entirety of his magnificent creation hovering in profile. He had been designing and building this flying machine for half a decade, toiling in the Arctic and in France, and now, even as her leather-colored skin was pasted dull against a heavy fog, she was indeed a splendid and spectacular marvel of technology. With tears wetting his eyes and cheeks, he strode back and climbed aboard. The crowd cheered and cheered, working themselves into a frenzy.

Just after 8:00 A.M., *America* was released from the *Olive*. Vaniman cranked up the larger engine and the great ship lumbered offshore, hovering at just one hundred feet, straining to bear the immense load it carried. Behind it trailed Vaniman's newfangled three-hundred-foot-long equilibrator. This one, comprised of metal gas canisters linked by steel cables, was connected to the lifeboat and served the dual purpose of guide rope and ground for the Marconi wireless outfit mounted in the lifeboat. Australian Jack Irwin, whom Wellman had hired "for his skill and pluck," and because of his long service as a wireless operator for the Marconi company, rode in the lifeboat, which hung suspended six feet below the steel nacelle. Crew members could climb into and out of the lifeboat via a rope ladder running between it and the nacelle—a daring maneuver while flying hundreds of feet in the air in heavy winds. As *America* turned east and rose higher into the sky, a *New York Times* reporter followed along in a chase boat. "The last glimpse of the airship," he reported, "disclosed the dim outlines of Mr. Wellman seated on a cracker box in the car, with navigator

Simon at the wheel." Wellman and the others lost sight of the hangar and crowd at the same time, "But we could hear them cheering somewhere back there in the gloom," he said.

Back on shore, as the crowd dispersed, Laura Wellman and Ida Vaniman faced a pack of eager reporters wanting to know their thoughts and feelings. Both were measured in their responses, upbeat and optimistic, neither betraying the fear and anxiety they surely felt. Laura Wellman's only concern was that her husband would be forced by weather or mechanical problems to abort the mission early and return to base, as had happened twice while he had been en route by airship to the North Pole. "All I hope is that they don't come back here," she said, gesturing to the hangar. "I have wanted them to be off for so long, I just dread to think of them coming back."

13

WELLMAN'S FINAL FLIGHT

AMERICA ENJOYED SMOOTH FLYING FOR THE FIRST FEW HOURS, cruising seventeen to twenty miles per hour and steering well. Jack Irwin sat at his wireless post in the forward compartment of the steel lifeboat, wireless receivers at his ears, wearing over them "thick woolen pads to drown out the whirr of the motor and propellers" so he could hear communications. He fired off a series of messages for Wellman, the first wireless messages ever sent from an aircraft. It was the modern media in the making.

WELLMAN'S WIRELESS MESSAGES TO THE TIMES
By Marconi Wireless to the *New York Times*, via Atlantic City
10:50 A.M.—
Good start. Everything working well, and have fresh north winds.
Fog still thick.—WELLMAN.
***

11 A.M.—
The sun is now coming out from the clouds.
Eight tanks of equilibrator above the sea.—WELLMAN.
***

11:15 A.M.—
Stopped motor to work wireless. Now going east-northeast.
Everything excellent.—WELLMAN.

***

12 NOON—

Am sending and receiving by Marconi while motors are going twenty knots an hour. Fresh north winds. Fog is still thick.—WELLMAN
***

12:10 P.M.—

Everything going splendidly. Making good progress.—WELLMAN
***

1:45 P.M.—

Sea smooth. Not crowding motors hard.

Averaging about 15 knots. All's going well.—WELLMAN

All day long crowds had gathered in Times Square in Manhattan, fixated on the updates that *The New York Times* was posting in virtual real time via electric bulletins. Onlookers gawked, awed by the new technology and the instantaneous news they were receiving, giving them the feeling that they were up there in the *America* with Wellman and his crew. When the wireless transmissions ceased around 5:00 P.M., news of a "tropical storm now stirring up the waves of the Gulf of Mexico" had everyone on edge.

Aloft in *America,* somewhere over the sea east of New York, weather was presently the least of their problems. One engine had failed, and the remaining working engine was now spitting fire. The exhaust pipes of the smaller Lorraine-Dietrich motor had become so hot that it began to "belch sparks . . . thick showers flying aft," threatening to ignite the cotton canvas covering of the steel car, or the outside of the balloon itself. As Wellman watched red hot embers flying from the exhaust pipes, he did not believe they would survive for long: "It seemed to me only a question of time when one of these fiery, incandescent masses . . . would set fire to the canvas and bring our little world to an end."

Vaniman told him it had been doing that all day, and looked worse than it really was, which did not reassure Wellman. Louis Loud saw the flames too: "When I look at those globs of fire, I don't feel good," he told Wellman. Vaniman explained to them that short of cutting the engine altogether, there was nothing they could do to fix the problem. Wellman and Loud watched the flaring "fireworks" from the exhaust for a few hours in horrified fascination. Eventually, Wellman realized that at least for now, the envelope and canvas nacelle covering were wet enough from fog and rain to prevent any fires. "I reduced the odds of eternity to 50 out of 100," he said, and they kept going with the single engine running, all praying silently to themselves.

Around 8:00 P.M. that night, after twelve hours of flying, one of the crew cried out an alarm warning from a speaking tube in the lookout of the lifeboat. Everyone looked, and all saw it: "Dead ahead, ghost-like in the fog, was a four-masted schooner." *America* was headed straight for the steamship, about to impale itself on one of the masts. Murray Simon, then at the wheel, "Jammed helm hard a-starboard," and they just barely cleared the ship as they watched men right below on the steamship's deck running wildly, waving their arms and shouting. *America* flew on. The captain of the steamship *Bullard* reported, "The airship had perfect steering . . . she came at us almost like the wind, and when almost on us she turned suddenly, like a motor car shooting around a corner . . . and passed harmlessly out to sea." They had avoided impact with the ship by only a few feet.

Wellman's plan from the start had been to follow the transatlantic steamer lanes, mainly for wireless communication of their progress, but also for safety, should they need assistance. Notwithstanding the near collision, at least now they knew they were on the steamer route. For the rest of the night, the single working engine propelled them forward, but gales began to pound them, driving the airship downward, so close to the ocean surface that whitecaps lashed against the nacelle and the lifeboat. All night they fought through the storm, and all through the next day. By Sunday afternoon, northeast of Nantucket Island, the weather calmed enough for Wellman to send a few wireless messages to the sponsoring papers. He reported that although they only had one engine and the winds were likely going to push them to the Azores, they were "keeping up the fight." Anxious readers, getting the news in the United States and London, cheered *America* on from their living rooms, breakfast tables, offices, and city streets. London bookmakers now posted even odds that Wellman's airship would pull off the crossing.

But winds conspired against them once more, driving hard from the northeast for most of the night. None of the crew had slept for two days, and then Vaniman, attempting to maintain proper pressure in the envelope by pumping the six ballonets, accidentally pulled the air valve rope instead of the hydrogen valve rope, and *America* soared upward to over three thousand feet. Everyone's ears rang and popped, and although Vaniman quickly released hydrogen and started their descent, they had lost a considerable amount of their irreplaceable hydrogen gas. Simon calculated their position and informed Wellman they were about 475 miles east of the coast of Maryland.

Checking the charts and assessing the winds, and factoring in the lost hydrogen, Wellman now believed that their best chance for survival was to try for Bermuda. That, or get low enough, and take to the twenty-seven-foot lifeboat, which was loaded with food and emergency provisions. The crew unanimously agreed to keep flying, though to a man they now knew they could not make it to Europe. Simon wrote in his journal, "The *America* will die from sheer exhaustion, a sort of bleeding to death, and before the last comes we must take to the boat."

It was now early Monday afternoon, and they had been airborne more than fifty hours, with virtually no sleep. Although they had agreed to try for Bermuda, it was only a tiny speck in the vast North Atlantic Ocean, and it was likely they would miss it. So, they began considering the dangerous task of disembarking *America* and loading onto the lifeboat while the airship hovered above them, an untried maneuver that would be tricky and not without risks. Someone would need to remain at the airship's helm until the last moment, steering straight into the wind to avoid being struck broadside and capsized by surging waves.

Wellman and his crew discussed their predicament. They had lost an estimated one-seventeenth of their hydrogen, and when the sun went down, the remaining gas would cool, possibly enough to plunge them into the ocean. Equally disconcerting, one of the last wireless messages they had received reported that the tropical storm that they had been guaranteed to miss was heading north, about to make landfall in Florida. If local winds kept driving them south, they would encounter the storm. Still, all agreed to fly on through the night.

As they puttered along through the darkness they ate as much as they could, "cold ham, ship's biscuits, tinned meats," and drank water. They had not stopped working for days, and now Wellman, up at the helm, nodded off, at times hallucinating in a sleep-deprived stupor. "My eyes began seeing things in the gleaming horizon or the gloomier depths covered by passing clouds," he wrote, "I saw a hundred steamers . . . trains of cars, rushing automobiles, tall buildings with shining lights. Then I shook myself, and saw nothing at all, only to drowse again, more optical delusions; then rouse, and nibble, and smoke."

Just before sunrise on Tuesday morning, Wellman believed he saw lights from a ship on the horizon. Shaking himself awake, he thought it must be another hallucination, so he took off his glasses, cleaned them, blinked his eyes a few times, then looked again. No, it was real. He yelled to Vaniman to make some kind of signal, and Vaniman wadded a bunch of gasoline-soaked rags, tied the bundle to the end of a long wire, then lit them and suspended them far below the lifeboat. All waited anxiously, and at last, the steamer changed course. It had seen them.

Irwin quickly tried the wireless but got no response. Using an electric "blinker" light, Irwin flashed Morse code, and the ship flashed Morse back. Irwin signaled that they had wireless, and within a few minutes, he was communicating with the steamer's operator. It was the SS *Trent*, bound for New York from Bermuda: DO YOU WANT OUR ASSISTANCE?

America: YES, COME AT ONCE. IN DISTRESS. WE ARE DRIFTING. NOT UNDER CONTROL.

The two parties agreed to convene at first light.

There was heated discussion among Wellman and the crew of just how to transfer from the moving airship, which was flying along at fifteen to twenty miles per hour, to a moving steamship, itself running at full speed to keep up. Vaniman, known for his daring high-wire acrobatics, suggested having the *Trent* sail right beneath them, and the *America*'s crew could slide down ropes to the deck below. This proposal was dismissed because of the danger of colliding with *Trent*'s masts or smokestacks and blowing them all up. They would have to risk deploying the lifeboat.

Just after sunrise, Irwin sent a wireless message to the *Trent*: WE ARE GOING TO LAUNCH THE BOAT. STAND BY TO PICK US UP.

By now, *America* was flying just above the steamship's rail, at a 45-degree angle, close enough that Wellman and his crew, having descended the ladder from the nacelle and boarded the lifeboat below, could see hundreds of passengers on the deck, waving and shouting and recording still- and moving-picture footage of the event. At that moment, the nose of *America* rose abruptly, nearly tossing Wellman and his crew from the lifeboat into the sea. Assistant engineer Aubert deduced the problem and acted fast, scrambling back up into the car and pumping air into one of the empty ballonets to even their trim just "before the ship stood on end and car and lifeboat tore loose." Then he climbed back down into the lifeboat.

Having leveled off, Vaniman pulled the rope controlling the hydrogen release valve, and they dropped down toward the water, with air and sea spray rushing around them. At five feet from water, they yanked the release hooks, and the lifeboat fell with a violent slap onto the ocean surface. *America,* relieved of a ton of weight, lofted and drifted away, dragging the equilibrator behind. The heavy metal snake struck the lifeboat's port bow and bashed a hole just above the waterline as it went. The vicious whipping end of the equilibrator also hit Loud and Aubert, injuring them with its final parting glance.

The *Trent* was instantly upon them, churning at fifteen knots, its giant bow about to impact them as Wellman's men grabbed for the oars and rowed for their lives. The *Trent* hit the small craft, half swamping the lifeboat, but somehow, they did not capsize. Shouting up to the deck, Wellman called for a line, which was thrown down. Everyone grasped for it, and Wellman caught hold, but it was moving so fast it seared his palms and fingers and he had to let go. As the *Trent* steamed by, the little lifeboat was sucked behind in a whirlpool and driven toward the propellers. Some of the men screamed, fearing being slashed by the blades, but the ship moved beyond them, where they bobbed along in its huge wake. Vaniman leaned over the side and vomited.

On the *Trent*'s third pass they caught and secured the line, and within minutes were ascending the steamer's rope ladder and boarding amid cheers, applause, and embraces. They were soaked, exhausted, bruised and bleeding, but alive.

And the *America* was gone.

14

A HERO'S JOURNEY

WALTER WELLMAN HAD FINALLY BECOME A HERO. HE HAD, INDEED, accomplished "the biggest thing in the world." By the time he and his men were plucked from the sea, the *America* had been flying for seventy-one and a half hours and traveled 1,008 miles, shattering every record for manned flight in a motor-powered dirigible. Count Zeppelin's previous record had been just thirty-seven hours. And though more than once the flight had nearly ended in calamity, it was by every measure a remarkable and unparalleled achievement.

On October 19, 1910, just past noon, the *Trent* steamed into New York Harbor carrying the rescued aeronauts. By now, Wellman had dispatched a four-page firsthand account of the harrowing and historic flight, published in *The New York Times,* and thousands of people anxiously awaited the arrival of Wellman and his crew, many onlookers crowding the wharves of the harbor's waterfront to get a glimpse of them. *The New York Times* and *The London Daily Telegraph,* cosponsors of the expedition, had choreographed the heroes' return expertly. They had chartered a tugboat to meet the *Trent* and then convey the airship crew to the wharf at 42nd Street. Aboard the tug rode Laura Wellman and daughters Ruth, Rae, Rita, and Rebecca; and Ida Vaniman.

The reunions were sweet and cinematic, with reporters placed by both papers to capture the first embrace between Wellman and Laura: she held his face in her palms and kissed him on one cheek, then the other. She was careful not to hug him too forcefully, since one of his arms was in a sling, injured during the hectic deploying of the lifeboat and the boarding of the *Trent*. Aubert's head was also wrapped in a bandage, and Louis Loud had a large, bluish purple knot rising on his forehead, but otherwise, the men appeared fine.

Wellman and his crew transferred to the chartered tugboat for a triumphant cruise through the harbor. Steamship captains blared their horns as they passed; ferry passengers rushed the rails to wave and call out well wishes and congratulations; small pleasure craft came alongside the tug to cheer and chat briefly with Wellman and Vaniman, doffing their hats to the men.

At the docks at 42nd Street, spectators blocked the roadway, many of them waving copies of *The New York Times* containing Wellman's dramatic dispatch, shouting "Bravo!" and "Well done!" Police hustled Wellman and his entourage of crew, wives, and family through the exultant throng to waiting taxis, which whisked them across to Midtown Manhattan for a welcome reception at the Times Building near Times Square. The ebullient ticker-tape reception rivaled that which the Apollo 11 astronauts would enjoy after their successful lunar landing and return to Earth in late July of 1969.

It was an overwhelming response, one that Wellman relished. Between the reception at the Times Building and a press conference later at the luxurious Waldorf-Astoria where Wellman and his family were staying, he had been handed scores of telegrams from friends and admirers, including one he particularly cherished:

I TAKE THIS OPPORTUNITY OF HEARTILY CONGRATULATING YOU UPON YOUR SAFE RETURN, AND YOUR MORE THAN INTERESTING EXPERIENCES. I HOPE TO BE ABLE TO HEAR ABOUT THEM FROM YOU YOURSELF, AND TO EXTEND MY CONGRATULATIONS IN PERSON; YOU HAVE PERFORMED A NOBLE FEAT. WITH HEARTY GOOD WISHES, FAITHFULLY YOURS,

THEODORE ROOSEVELT

Wellman also received a telegram from his longtime rival that he found gratifying: "Hearty congratulations. You have done a grand and noble thing. Peary."

For the next months, Wellman, Vaniman, and the crew were glorified and heralded as heroes. Most of the press, and particularly those in the United States and in England, showered Wellman with praise, in lengthy glowing stories with headlines like these: "Wellman Has Made Good"; "Wellman's Trip Proved Grit"; "Wellman Is Great"; and "Daylong Crowd at Wellman's Door."

Naysayers and doubters remained, but in the main Wellman was lauded for his courage and tenacity. Wellman mostly ignored the negative coverage and reveled for a time in what he had accomplished, at the part he had played in the dawn of the aerial age. He had time to reflect on the fate of the *America*, which had borne him above the polar seas of the high Arctic and across a thousand-mile expanse of the Atlantic Ocean. He recalled that when they had finally released the lifeboat from *America* and his dirigible had floated off into the sky, both he and Vaniman had tears in their eyes:

> The last we saw of our good airship . . . she was floating about 800 feet high, 360 statute miles east of Cape Hatteras. A day or two later, in all probability she disappeared beneath the waves . . . Good old *America,* farewell. Thank you for the noble comrades and rare experiences you have brought me, for the lessons you have taught us. You played your part in the game of progress. In the years to come many aircraft will cross the Atlantic, and you will be honored as the ship that showed the way.

Walter Wellman was right that in the coming years, many aircraft would indeed cross the Atlantic Ocean, and the first would be a dirigible, just as he had predicted. But it would not be Wellman at the helm. Despite entreaties and financial offers for him to build another airship and try again, Wellman was finished with his airship experimentation and exploration. Rumors swirled that he and Vaniman had had a falling out, though neither man confirmed or denied the gossip, and Wellman, in articles he wrote soon after the historic flight, heaped high praise on Vaniman as an heroic team member.

Though Wellman never stated explicitly his reasons for quitting aerial exploration, he certainly would have contemplated his numerous scrapes with death on an airship—twice over the rugged mountains and seas above Spitsbergen, and once over the Atlantic. Each time, by some combination of luck and providence, he had escaped with his life. Perhaps his intuition told him that there was only so much luck one man was allotted. And Wellman, as always, had numerous other obligations, many of them financial. He had three families (that we know of) to support and had determined to write a memoir—*The Aerial Age*—which he wrote between the last flight of the *America* in October of 1910 and its publication before the end of 1911. He also briefly parlayed his celebrity status into a popular lecture tour that year, retelling his many adventures.

As a result, when Melvin Vaniman announced in late 1911 that he was building a new airship called the *Akron* at the Goodyear factory in Akron, Ohio, Walter Wellman was not involved. Six months later, at 6:15 A.M. on July 2, 1912, Vaniman and a crew of four took off from Atlantic City into a brilliant sunrise, and the *Akron* rose briefly above a large gathering of onlookers assembled along the boardwalk. Ida Vaniman was among the spectators—as were the wives of three of the crewman. *Akron* ascended to about 2,500 feet, then suddenly pitched forward pointing directly downward, as if attempting a fast descent. Moments later, to everyone's horror, the envelope burst, ripping away from the car. The car—in which Vaniman and the crew rode—spun and twisted as it plunged from the sky, spiraling violently into the shoal waters just off the shore of Absecon Bay. The burst envelope fell nearby, then drifted away on the tide.

There were no survivors.

In the days that followed, after search and rescue teams had recovered and identified the bodies, which had been "driven into the sand beneath the car," Walter Wellman expressed his condolences and praise for Vaniman in the 1912 Fourth of July's *New York Times*:

Melvin Vaniman was one of the bravest and most resourceful men I have ever known. There are two fine things in this world—courage and brains. Vaniman had both . . . The few of us who have sailed the air with him know him at his true worth. His death makes our hearts heavy. It is not only the loss of a brave man, but of one who with his valor, his persistency, his marvelous ingenuity as a mechanic, seemed destined to do great things . . . He had set his heart upon crossing the Atlantic by airship. With life and time, I think he would have achieved it.

Walter Wellman never flew again, either in an airship or an airplane. He had experienced enough airborne adventure for one lifetime. Undoubtedly, Vaniman's death was part of the reason, as Wellman could very easily have been on the *Akron,* right there alongside his good friend and trusted airship designer. But he would keep writing about the beguiling and deadly "aerial age," in which he had played a part. In the years that followed, the remarkable Walter Wellman wrote prodigiously about many aspects of flight and, in particular, of his own airship adventures and explorations, publishing his recollections, opinions, and predictions in the leading scientific and geographical journals of the day, as well as in more "popular" general magazines. He remained, for a time, one of the most respected and highest paid journalists in the United States.

He would keep a watchful eye on the future development of the airship both as a potential tool for commercial travel (about which he was dubious) but more important, as the best method by which to reach the North Pole and explore the immense polar sea beyond the pole, which remained yet uncharted. Although Robert Peary's polar claim had been validated and generally accepted, doubts still lingered, particularly because Peary's logs and diaries had been sealed by the National Geographic Society and were available only to Peary and his family.

These doubts, Wellman knew, would continue to lure men to the North Pole, in the event that Peary was one day proved a fraud and the prize could be theirs instead.

As far back as 1893, Walter Wellman had predicted that "aerial navigation will solve the mystery of the North Pole and the frozen ocean." He
never wavered in that belief. "Flight cannot be stopped," he wrote. "It is in
the blood of man to conquer the air, and he will go on till he does it, no
matter how long it takes or how great the cost."

Whether Wellman would be alive to see it happen was uncertain,
but he would remain an active observer and commentator, curious to see
which brave explorers would be the ones that followed his pioneering path
and fulfilled his dream.

PART TWO

THE WHITE EAGLE OF NORWAY

Roald Amundsen and the *Norge*

People who ever knew Amundsen always tell you first about his re-markable eyes. His long head and hooked nose gave him the look of an eagle, an effect which his imperial white moustache only accented. Yet his eyes were his most arresting feature. Years spent on the decks of vessels and amid limitless sweeps of ice and snow had given them a chronic squint. Through narrowed lids peered those grey eyes, boring through one as their gaze passed on into infinite distances.

—Lincoln Ellsworth, *Beyond Horizons*

15

THE END OF THE HEROIC AGE OF EXPLORATION—THE BIRTH OF THE GOLDEN AGE OF FLIGHT

AT NEARLY THE SAME TIME WALTER WELLMAN WAS HANGING UP his aviator cap and goggles in 1911, determined never to leave the Earth's surface again by either airship or airplane, Norwegian explorer Roald Amundsen began taking flying lessons. Like Wellman, Amundsen knew that the Heroic Age of Exploration was at an end, even if his participation in that now extinct epoch had brought him fame and fortune. "Aircraft has supplanted the dog," he wrote without nostalgia or regret. "The future of Polar exploration lies in the air."

Amundsen had to concede that the traditional dog and sledge had served him well, as had the sailing vessel. In 1903–06, aboard his specially-built sloop *Gjøa*, Amundsen and his small crew successfully navigated the entire Northwest Passage—the first in history to accomplish it. In 1911, Amundsen, still slogging away behind sled dog teams, had been first to reach the South Pole, a stunning achievement which resulted in many experts and layfolk alike crowning him the greatest polar explorer in the world. But Amundsen now understood that, in his own words, "the dog and the sledge have wholly outlived their usefulness in the Arctic as instruments of exploration. Their place now, though forever glorious, is in the museum and the history books."

Amundsen knew full well that his name and exploits would comprise major chapters in those history books, but henceforth, the stories of his part in future exploration and discovery would be connected to flight over the polar regions. The new pioneers of the poles must take to the air. He went on to boast, "I make bold to claim for myself the distinction of being the first serious Polar explorer to realize this fact, and the first to give a practical demonstration of its future possibilities." Walter Wellman would certainly have disagreed, having only recently flown an airship over a thousand miles, clearly demonstrating its potential for the very kind of long-range flight necessary for polar exploration. Amundsen's slight of Wellman appears rooted in his use of the words "the first *serious* Polar explorer," a class to which he did not believe Wellman belonged. Amundsen thought of Wellman as merely a showman. That was unfair and uncharitable. It's true Wellman managed to garner global publicity of his exploration by being a great promoter and a widely read journalist, but he still achieved numerous exploratory firsts. By any metric, Wellman showed vision, courage, and grit, and he was trying to advance science and mankind's understanding of the world through his airship adventures.

Whatever the case, Wellman would no longer be flying anywhere, while Amundsen believed he personally had much more to accomplish. In 1913, Amundsen had been in Europe when he first watched airplanes in flight and began to contemplate their potential for polar exploration. In much the way Wellman had stood on the rupturing, moving ice and realized that the airship would make things much easier, so did Amundsen experience an epiphany. Watching those planes fly—crude as they were at the time—Amundsen was transfixed. "I stood with fresh memories of the long sledge journeys in Antarctica," he remembered of his painful South Pole journey, "and watched the machine in the air cover distances in one hour that would have taken days and cost fearful effort in the polar regions."

For Amundsen, it was a matter of when, not if, he would take to the skies toward the poles. He would need to choose the right time, the right goal, and a much technologically improved flying machine, but as early as 1913, the dream had embedded in his imagination. With the outbreak of World War I the following year, that dream was put on hold and would have to wait—for about a decade, as it turned out.

Roald Engelbregt Gravning Amundsen was born in Borge, Norway, on July 16, 1872, the fourth son and youngest child of a prominent family of shipowners and sea captains. Amundsen seemed destined for a life as a seafarer—as many of the men in his family were—but in his early years he promised his mother, Gustava (his father, Jens, died when he was just fourteen), that he would eventually go to university and become a doctor. It was a promise he tried to keep, though from boyhood he had been enamored with the outdoors and the adventurous life, spending hours skiing, snowshoeing, climbing hills and mountains. He was inexorably drawn to snow and ice.

He was also keenly aware of the national glories recently heaped on his elder countryman Fridtjof Nansen, who in 1889—when Amundsen was just seventeen—returned a conquering hero after becoming the first to cross Greenland. In true Norwegian style, Nansen had done it on skis, and his return up the Oslo fjord to Christiania was celebrated with tremendous fanfare, with pleasure boats following along waving flags and blaring their horns. The city was shut down by thousands of cheering admirers who lined the shoreline and streets to see Nansen and his five companions arrive. But it was really Nansen that everyone wanted to see, including the teenage Roald Amundsen, who was among those spectators bursting with national pride. It was Nansen, Amundsen would write, "Whose daring exploits . . . made him the idol of my boyhood."

A poor student, Amundsen's mind was always elsewhere, always focused on adventure. He passed his college entrance examinations—if just barely—and in 1890, enrolled in the Royal Norwegian Frederick University in Christiania (Oslo). Amundsen floundered as a university student. He was often truant, preferring to be outdoors, and rather than studying his medical textbooks, he read the books of the British polar explorer Sir John Franklin.

Though Amundsen did not tell his mother, he was failing university, his attention distracted by Franklin's amazing tales of the polar region and the privations they wrought on those brave enough to venture there. "Strangely enough," Amundsen wrote of reading Franklin, "the thing . . . that appealed to me most strongly was the sufferings he and his men endured. A strange ambition burned in me to endure those same sufferings . . . in the frozen North." Amundsen credited the descriptions in Franklin's narratives as "decisive in the direction I chose to take through life."

Circumstances prevented Amundsen from having to tell his mother he was failing medical school: she contracted pneumonia and died when he was twenty-one and still enrolled. "Her death," he wrote wistfully, "saved her from the sad discovery which she otherwise would have made, that my own ambitions lay in another direction and that I had made but poor progress in realizing hers." He left the university immediately, and with a sense of newfound freedom, exclaimed, "I irretrievably decided to be an Arctic explorer." It was a life he had secretly been preparing for since his early teens, always sleeping with his bedroom windows wide open, especially during the coldest winter nights, as part of what he called a "conscientious hardening process." He would become tough, and he would learn to relish suffering.

The death of his loving mother was difficult for Amundsen, but the considerable sum of money he inherited gave him the freedom, as he put it, to "throw myself whole-heartedly into the dream of my life." He trained vigorously, organizing weeklong ski tours with just one or two companions— sometimes his older brother Leon—and was keen to experience that suffering he had sought. On one trip, they became lost in a blizzard and nearly froze to death in their reindeer-fur sleeping bags as the temperatures dropped to minus 40 degrees Fahrenheit. But being lost, frozen, and famished was necessary training for the Arctic, and he continued skiing and hiking in the mountains rising above Oslo, hardening his body and mind for future endeavors. He also completed his obligatory military training, spending the required seven months in the army: "This I was eager to do," he wrote, "both because I wanted to be a good citizen, and because I felt

that military training would be of great benefit to me as further preparation . . ."

Amundsen knew that the life of a polar explorer required seamanship, so he hired onto a commercial sealing vessel working Norway's high
northern waters. He spent the next few years mostly aboard ships in the
Arctic Ocean off Greenland and Iceland, and then on merchant vessels—
traveling as far as the Americas and West Africa. Amundsen excelled at
studies related to his own goals and subjects he cared about, and in 1895,
he achieved his mate's certificate, qualifying him to apply for a position as
mate on a polar voyage. He continued to read voraciously on polar exploration, and he came to believe that a central flaw in most failed expeditions
was a division of leadership between the expedition commander and his
scientific staff and the ship's captain and his crew. Amundsen vowed to
solve this common fracture, this "incessant friction, divided counsel, and
lowered morale." When he ultimately would lead a polar expedition, he
would be both expedition commander and captain of the ship, assuming
supreme control of all decisions and bearing the responsibility of the outcome.

By the time Amundsen was twenty-four years old, he had spent enough
time aboard ships to qualify as a master mariner in Norwegian maritime
waters. He was by then a superb physical specimen, lean and sinewy yet
tautly muscled, and also having developed the endurance needed for days
on end of constant work and exertion, whether on a ship or in the mountains. In 1897, he learned of a Belgian voyage to Antarctica in search of
the magnetic South Pole and immediately applied, even offering to serve
without pay—since at the time he was still in good financial shape from
his inheritance. Amundsen was accepted as first mate, and he spent the
next two and a half years as part of the Belgica Antarctic Expedition. As it
happened, the physician of the international scientific staff was none other
than Dr. Frederick Cook, who would eventually become deeply embroiled
in the controversies surrounding his McKinley summit and North Pole
claims.

The *Belgica* experience was a true polar apprenticeship in every way. The ship became icebound in the dense pack ice of the Bellinghausen Sea, and the crew endured the trials of Antarctic night—months of near-total darkness. Many of the crew succumbed to scurvy, others descended into depressed psychosis from the monotony and lack of light. It became, according to one recent chronicler of the events, a "madhouse at the end of the earth."[1] When the ship's captain became so debilitated from scurvy that he could no longer perform his duties, Amundsen, aided by Cook as physician, essentially assumed control of the ship. Amundsen and Cook spent many days together on the ice, exploring but mostly hunting penguins to feed the sickly crewmembers. Cook had determined that the rampant scurvy could be treated with half-raw penguin meat, though he did not understand exactly why. But he had deduced that those who ate the fresh uncooked penguin meat were improving, those who didn't got worse.[2]

Amundsen, Cook, and the rest of the crew survived for thirteen months aboard the ship encased in ice, and some of the sailors, as Amundsen put it, "went insane." Finally, with the *Belgica*'s hull leaking, Cook and Amundsen realized they must do something, or they would all surely die. The sun had by now returned and Cook noticed a pool of open water some seven hundred yards ahead of the ship. He believed that with the coming ice breakup, that basin of water—if they could get there—might lead them eventually to the open sea. He proposed a bold plan: using hand saws, pickaxes, shovels, and explosives, they would cut a canal through the ice all the way to the pool.

---

1 For an excellent, riveting book about the *Belgica* expedition, I highly recommend Julian Sancton's *Madhouse at the End of the Earth: The Belgica's Journey into the Dark Antarctic Night* (Crown, 2021).

2 At the time, it was not yet fully understood that scurvy was caused by Vitamin C deficiency, and that nearly all mammals produce their own—in quantities sufficient to prevent or even treat scurvy. Scottish surgeon James Lind conducted a controlled experiment in 1747, which strongly suggested that consuming citrus fruits or juices remedied scurvy, though he still did not understand exactly why. When plentiful citrus fruits or other fresh fruits and vegetables high in Vitamin C were not readily available, fresh (and especially uncooked or undercooked) meat worked well.

Amundsen and Cook directed the emaciated men for many weeks of backbreaking, freezing work cutting through the ice beyond the bow of the *Belgica,* then using dynamite to dislodge the weakened sections. After many weeks of toil, they had created a channel just wide enough for the ship to pass through. Using dynamite dangerously close to the hull, they blasted the *Belgica* free, fired up the engines, and steamed the ship through the narrow ice canal. "Joy restored our energy," wrote Amundsen, and with all speed we made our way to the open sea and safety." Amundsen returned home to Norway a seasoned polar explorer, his appetite whetted for more. He had learned how to travel across deadly crevasses, how to navigate through broken and ruptured terrain, and most importantly, how to lead men. He vowed to never participate in another expedition in which he was not the expedition leader.

That drive to leadership served Amundsen well. Over the next two decades, employing small crews and a severe but fair leadership style, he successfully navigated the *Gjøa* through the Northwest Passage in 1906, and in 1911 emerged triumphant from his race with Sir Robert Falcon Scott as first to the South Pole. He was by then arguably the world's greatest living explorer, eclipsing even his childhood idol and countryman Fridtjof Nansen, who had retired from exploring and had turned his attentions to global humanitarian work. But Amundsen wasn't finished. In his mind, there remained more exploring to do.

In 1924—at the age of fifty-three—Roald Amundsen surprised the world with a stunning announcement: He would fly an airplane from Spitsbergen to the North Pole, and then onward to Alaska, "from continent to continent across the Polar Sea," a distance of about two thousand miles. No such thing had ever been attempted in an airplane.

It was an audacious idea, bordering on insane. But the Golden Age of Flight had arrived, and Roald Amundsen, by now being called "The White Eagle of Norway" for his shock of white hair, his sharp, beaked nose, and his piercing gray eyes, was going to try.

16

"A RACE WITH DEATH"

ROALD AMUNDSEN STOOD ON THE SNOWY SHORELINE OF KINGS BAY (Ny-Ålesund), Spitsbergen, a mining settlement about sixty miles due south of Danes Island. It was the afternoon of May 21, 1925, and he eyed the two Dornier-Wals "flying boats" poised on the snow- and ice-covered bank, down which the airplanes would taxi to the fjord ice for takeoff. With Amundsen was forty-five-year-old American Lincoln Ellsworth, whose millionaire father had provided the funds to build the two state-of-the-art airplanes that stood ready for departure. The Arctic sky was pale and clear, and a light breeze fanned the fjord.

Amundsen watched as a few dozen ground crew pushed the airplanes, *N24* and *N25*, down the slick ice ramps to the frozen fjord. The airplanes were skidded on their duralumin, whale-like fuselages, for they had no landing gear, and were designed to be able to take off from water, snow, or ice. Each weighed four tons, was fifty-two feet long, and had a sixty-eight-foot wingspan. As the pilots climbed into their cockpits, Amundsen listened to the rumble of the 360-horsepower Rolls-Royce engines warming up, and he knew it was time to go.

Amundsen and Ellsworth, bundled in thick fleece leather jackets, parkas, fur hats, and gloves, buckled their parachutes tight and strode down to the airplanes. Amundsen climbed in and took his seat in the first open cockpit of *N25,* beside his pilot; Ellsworth, also as navigator, sat next to his pilot in the *N24*. A mechanic rode in each of the aircraft, behind the pilot and navigator.

N25 roared away first, the powerful engine pushing 2,000 revolutions per minute as the airplane sped across the ice toward the mountains and glaciers beyond, rising from the ice in a trail of swirling spindrift. *N24* followed, cracking the ice and ripping rivets from its fuselage as it tore along the uneven surface, according to Ellsworth, "with the swiftly gathered speed of an arrow lunged forth after *N25* toward that distant glacier." The American Ellsworth was awed by the spectacle as the settlement at Kings Bay receded fast from view, "the dark ring of spectators and the coal station, its shops and sheds, black against the snow . . . dwindling to the size of huts." Up the aircraft went, lifting and banking hard. "I felt like a god," remembered Ellsworth, "and I found myself looking over the wall of the fjord into the North—looking over a fleecy sea of fog."

The Amundsen-Ellsworth Expedition was officially underway. Traveling 80 to 90 miles per hour at up to ten thousand feet of altitude, if all went well, Amundsen could be landing at the North Pole by midnight.

Amundsen was fortunate to be aloft above the mountains of Spitsbergen at all. During the past few years, he had reached a low point in his life. Despite all of his fame, and decades of successful fundraising to finance his expensive expeditions, bad business dealings had left him woefully in debt to the point of bankruptcy. In 1924, Amundsen had traveled to the United States on a lecture and writing tour hoping to recoup some of his losses, but they had not gone as well as he had hoped. The articles he wrote for U.S. newspapers garnered little income, and the lecture tour, as he put it, was "a financial failure."

In early October that year, he had sat alone in his hotel room, deeply depressed. "It seemed to me," he wrote of that dark time, "that my career as an explorer had come to an inglorious end . . . I was nearer to black despair than ever before." As he sat pondering his uncertain future, the phone in his room rang. Worried that it was likely creditors, he almost didn't pick up the phone. But something compelled him to answer. The voice on the other end of the line was Lincoln Ellsworth's.

"I met you several years ago during the war, in France," Ellsworth began. Amundsen winced. He had met hundreds, even thousands of admirers over the years. The last thing he needed now was idle chit-chat with a fan with whom he had once shaken hands. But for some reason, Amundsen didn't hang up. He kept listening. "I am an amateur interested in exploration," Ellsworth went on, "and I might be able to supply some money for another expedition."

That got Amundsen's attention. He invited the younger man to come to his room immediately so they could talk. Amundsen encountered an inquisitive and ambitious man with tightly cropped hair and a beaming, clean-shaven face that was chiseled and handsome as a movie star's. Though Ellsworth admitted his previous exploration was limited to having led a trans-Andean expedition from the Amazon Basin to the shores of the Pacific Ocean in Peru, he highlighted that he had trained as an aviator, serving in World War I for the U.S. Army. He was also a surveyor and engineer, having studied science at Yale and Columbia universities. Ellsworth told Amundsen he had "an independent income and a strong thirst for adventure," which, to Amundsen at the time, was a perfect combination. But there was a catch to the money—as usual. Though Ellsworth had some independent financing (in truth, only about ten thousand dollars readily available)—they would need to obtain the bulk of the necessary sum from his ailing father, the wealthy industrialist and coal magnate James Ellsworth.

Amundsen agreed to meet with Ellsworth's father at his estate in Ohio. There, summoning a combination of stoic charm and captivating storytelling of his exploits, he had come away with a pledge of $85,000, enough to finance the building of the amphibious airplanes he would require for the trans-polar flight. Amundsen estimated the total cost would be $150,000, but he was confident he could secure the difference from the Aero Club of

Norway, so he shook hands, had contracts drafted, and sealed the partnership with Lincoln Ellsworth.

The first few hours, flying abreast above fog and clouds north of Spitsbergen, were magnificent. Bouncing along in light turbulence, they flew directly into a double solar halo, "two wraithlike circles of rainbow with reversed colors." Ellsworth and Amundsen waved and signaled back and forth to one another to communicate because the radios they had ordered had never made it to Kings Bay. Though their faces were freezing, both beamed with exhilaration as they climbed higher, up to nearly ten thousand feet of altitude. Below, the fog diffused, revealing the seemingly infinite plain of ice pack. As Amundsen stared ahead, trying to take a bearing from the sun, he could discern no horizon. "Sky and ice blended into one," he wrote. He was dazzled by the spectacle, the immensity, and humbled, too, thinking of all of the explorers who had trudged across the "vast white waste," and the many who had given their lives there.

In eight hours, they had flown a distance that Amundsen estimated to be approximately six hundred miles, and he motioned for his pilot to descend. They spiraled downward to begin scouting a place to land $N25$. $N24$ followed. The lower they flew, the more treacherous and ominous the polar pack appeared: upthrust pressure ridges, deep fissures of broken ice, circuitous open water leads pocked with giant icebergs. Landing the seaplanes in open water was safest, but once accomplished there was the chance that the ice could close in around them and crush the flying boats. Landing on ice presented its own hazards, since nowhere was flat and smooth, and the craft could easily be damaged, or could even crash as it skipped along on its aluminum underbelly.

As Amundsen scanned the surface and pondered options, one of the engines started misfiring, and his pilot, Hjalmar Riiser-Larsen—a skilled Norwegian naval airman—dropped $N25$ fast, navigating a forced landing into a narrow lane of water filled with slurry and pancake ice and lined on both sides by tall icebergs. As the aft engine coughed and sputtered,

Riiser-Larsen coaxed *N25* deftly into the ice canyon and touched the belly down onto the slushy open water. They hydroplaned and corkscrewed through the frozen slosh, finally coming to a stop with the airplane's nose pressed right up against a large iceberg. As Amundsen got out to take bearings and figure out what to do next, he looked up into the sky, hoping to spot *N24* circling around, but he could not locate it nor hear it.

Aboard *N24,* Ellsworth had seen *N25* spiraling down in emergency descent, and his own pilot, Leif Dietrichson, had immediately started looking for a better place to land. It took ten minutes before they saw a "tiny lagoon that was ice free," and Dietrichson set the plane down into water, barely missing "bergs of old blue Arctic ice, twenty, thirty, and even forty feet thick," skidding and finally colliding nose-first into a wall of ice at the end of the open lead. As the engines cut and all grew quiet, Ellsworth looked around, stunned by the "white and blue chaos" of the water, the ice floe walls, and the tremendous icebergs floating nearby. A lone seal surfaced right next to them, watching them curiously, when Ellsworth heard a warning bell ringing and then Dietrichson yelling, "The plane is leaking like hell!"

As the mechanic began furiously pumping water from the leaking fuselage, Dietrichson and Ellsworth clambered out of *N24* onto an ice floe covered in two feet of snow. Dietrichson cussed when he saw that one of the engines had been damaged during their landing. Since it was obvious that they were going to be there for some time, they unloaded all heavy gear from *N24* to keep it from sinking, then set up a tent and raised a Norwegian flag from a pole on the highest nearby point, hoping the *N25* would see them. A sextant reading determined they were at 87° 43′ north, about 150 miles short of the North Pole. That was unfortunate; they must have been blown toward the west. But their greatest immediate concern was trying to locate *N25* and its crew. "In the utter silence," Ellsworth observed, "this place seemed to me to be the kingdom of death."

In the morning, Ellsworth and Dietrichson put on skis and climbed a high, steep pressure ridge and glassed the ruptured expanse with binoculars. There, two to three miles away, Ellsworth spotted a wing of the

N25, and after a time, the outline of three figures moving around near the plane. Amundsen had seen them, too, and since they had no radios, eventually the two groups began communicating through a combination of rudimentary semaphore and Morse code, waving flags back and forth to one another as each group peered through field glasses and jotted down the simple messages. Ellsworth managed to convey that *N24* was leaking and had sustained engine damage. Amundsen reported *N25* undamaged (his mechanic had already worked on the engine issue that had forced their landing in the first place).

Amundsen realized the gravity of their situation immediately. They had, it appeared, just one working aircraft, though it was currently floating in a narrow lead and in danger of being crushed. There were six men a few miles apart, separated by moving ice and long, upthrust pressure ridges. They had about a three weeks' supply of food and were about six hundred miles north of Spitsbergen. The northern tip of Greenland was at least four hundred miles to the southwest. Without dogs, he knew that attempting to walk back south to civilization meant certain death. "The only prospect of salvation," he determined, "lay in transferring the whole party to the *N25* and making every effort to get that one plane in the air again."

But first the two groups needed to be reunited. Over at *N24,* after a few days Ellsworth had decided that he, Dietrichson, and the mechanic, Oskar Omdal, needed to make it to Amundsen to see what plan the expedition leader had come up with. They secured their camp and provisions and started out on skis across the shattering, and shifting ice that was covered with two to three feet of snow. Laden with heavy packs, they made slow progress, though Ellsworth did notice that the floe they were crossing seemed to be drifting in the direction of Amundsen's group, which was encouraging. But a large lead of open water separated them from *N25,* so they skirted around searching for narrow places or ice bridges to cross.

They eventually found a recently frozen section that appeared crossable. It looked precarious, but they were committed, and they took off their skis for the maneuver. Omdal moved to the front, with Ellsworth and

Dietrichson spread out behind to disperse their weight as they crept along. In the distance, they could see the outlines of two of the men from *N25* trudging their way toward them. There was a loud crack, like a rifle shot, and suddenly the ice collapsed beneath them, plunging Omdal and Dietrichson into the freezing seawater. Ellsworth managed to leap onto a floating ice cake and extend his skis toward Dietrichson, who was flailing and gasping for air. Dietrichson caught the end of a ski tip, and Ellsworth pulled him near enough to grab onto one of his pack straps and haul him out of the water onto the wobbly ice cake, where Dietrichson lay splayed out, hyperventilating.

Ellsworth turned to see Omdal, with only his head above water, gasping for help. Ellsworth crawled to the edge of the ice and again reached out both of his skis. Just before going under completely, Omdal lurched and grasped, catching hold of the end of one ski. "I pulled him through breaking ice until his pack was in reach," remembered Ellsworth. "Then, partly supported by the firm ice, I could do no more and merely held on." In the torrent of moving floes, Omdal's legs were pulled under the ice, but Dietrichson had recovered enough to bear-crawl over to help. Omdal's heavy rucksack was still dragging him under, but they yanked him close enough to cut the straps and pull him back onto the ice. He was barely breathing and hardly conscious. Blood was pouring from his mouth, and "he had broken off five front teeth on the ski points and ice edges," Ellsworth wrote. He was hypothermic, but he was alive.

The three men, soaked and exhausted from the ordeal, knew they had to move or they would freeze. They slinked on hands and knees across the ice, using Ellsworth's skis—the others had lost theirs in the water—to crawl to a larger, firmer ice floe ahead. Once there, they moved through waist-deep snow over rolling hummocks until they met up with Amundsen and Riiser-Larsen, who had heard their cries and were hurrying toward them. Amundsen and Riiser-Larsen took the two remaining rucksacks and guided their soaking, half-frozen compatriots to *N25*, where they quickly got them warm clothes and wrapped them in blankets in the cockpit of

N25. The shivering men were also plied with cups of hot chocolate and whiskey.

Once they were revived, Amundsen addressed their predicament and stated his plans. Their only hope was to free *N25* from the ice chasm in which it was trapped, then somehow carve out a runway on the polar ice long and smooth enough for them all to take off. With half-rations, they had about a month's worth of food, and no more. By June 15, the day they would run out of food, they must either fly away or begin plodding into the vast whiteness toward Greenland. It was dire, he told them, "Truly a race with death."

17

"YOU'RE SUPPOSED TO BE DEAD."

DRAWING ON DECADES OF ARCTIC LEADERSHIP, AMUNDSEN IMME-diately broke the men into teams and outlined duties that they would perform in shifts, working round-the-clock in the twenty-four-hour daylight of the midnight sun. They needed to free *N25* from the arm of slush water it had landed in, which was now freezing and knitting together. If they didn't get *N25* out soon, it would be crushed. Dietrichson had reported that *N24* was too damaged to fly again, but they needed to return to the airplane to retrieve tools, any food, and fuel they would need in the *N25*. Last and most arduous, they somehow had to build a runway.

Amundsen set forth an organized, orderly schedule. Taking shifts, teams would work, eat and drink, and rest or sleep. Amundsen also wanted some scientific observations recorded—ocean depths, temperatures, and sea current and ice floe drift rates—which would be important records to have should they survive.

The first job took all six men at once. Using haul ropes as Riiser-Larsen gunned the engines, they managed to turn the airplane around and drive it up onto relatively flat, solid ice so that it appeared, at least for now, safe from being crushed by the shifting floes. Ellsworth, Dietrichson, and Omdal made two trips back over to *N24*, moving slowly and timidly given their recent hairbreadth escape. Fortunately, winds and currents had moved *N24* closer to them, and the ice had firmed up. They salvaged some remaining provisions and a full gas tank, dragging it back on a sledge.

For the next few weeks, all of their efforts were devoted to the runway. They had few useful tools, only a hatchet and three shovels, but they made ice chippers by strapping long-bladed sheath knives to the ends of ski poles, and they used their ice anchor as a bludgeoning club to break up large chunks of ice. The work was cold, grueling, and on their short rations, exhausting. They were eating only pemmican soup, chocolate, and biscuits, and more than once Ellsworth cursed himself for not shooting the seal that he had seen when they first landed. They had not spotted any since. The men became gaunt and feeble, hacking and chipping away and shoveling ever-shifting ice. There were petty squabbles as they camped in this distant, forlorn world, some shivering in tents and others hunkered down inside the *N25*'s fuselage to escape the constant, freezing wind. A couple of the men became depressed and even delirious, looking up to the wide skies, sometimes imagining they could hear the drone of rescue planes searching for them, but it was only the wind hurtling over ridges and through icebergs.

After three weeks, by June 14, 1925, they had moved what Amundsen estimated was a few hundred tons of ice. He well remembered the ordeal of cutting and chipping and exploding the Antarctic ice to free the *Belgica,* and this runway job was even more difficult as they had no explosives, and few tools. For the last few days, all six men had been working together boot packing the runway as flat as they could make it, stomping the hard surface hour after hour, day after day, until they had a relatively level lane about 450 yards long, just barely enough, the pilots estimated, to take off. The following morning was the day agreed upon to depart, and no one slept well that night, despite their exhausted and famished condition.

They rose early and assessed the ice field. It was hard and wind shorn, and Amundsen knew it was their only chance. Amundsen walked the entire runway, lining its edges with black cans as markers to guide the pilot in the monochromatic light. He noticed a crack running straight across the track that worried him; it was a few feet wide, "filled with water and mush," and he feared the shifting ice might split it open wide if they waited too long. At the end of the runway was a twenty-five-foot-high ridge they must clear, so he ordered the cockpit and fuselage stripped, and everything extraneous jettisoned from the *N25*, keeping only the barest and lightest necessities: clothes, a tent, a cookstove, and what little food there was remaining. In a heap on the ice, they had discarded two sledges, a canoe, rifles, shotguns, heavy metal ammunition cans, skis and ski boots, and an expensive movie camera.

Under low and ominous cloud cover and light fog, four of the men crammed into the rear compartments of the Dornier-Wal, and Amundsen took his place up in the cockpit with Riiser-Larsen, who would pilot the aircraft. Amundsen wasn't certain they had enough fuel to reach Spitsbergen even if they managed a takeoff; it would be close. Ellsworth, harboring "a superstitious notion that it was bad luck to watch," crawled into the dark tail end of the airplane where he could not see out. Riiser-Larsen fired up the engines and the airplane vibrated as he drove forward at full throttle.

The flying boat roared down the ice-way, the hull scraping and grating. It skipped along with increased speed: forty, fifty, then sixty miles per hour as they cleared the big crack and the nose lifted, and they were airborne. Ahead loomed the high ridge, and as everyone tensed and braced for impact, *N25* rose and skimmed over it, the hull clearing by inches. The men cheered, and they banked south, watching the dilapidated *N24* diminish in the chaotic jumble of ice and snow below.

They flew blind through "thick fog, dark and clammy," recalled Ellsworth, who had finally come forward to peer through a porthole window. Hour after hour, the engines droned as they pushed against a headwind through intermittent haze and sunshine. Sometimes Riiser-Larsen took them very low, just a few hundred feet above the pack, buzzing over the tops of icebergs as Amundsen watched the fuel gauge nervously. To reward the men for their labors, Amundsen distributed a carton of chocolate bars, and the men scarfed them down. After eight and a half hours of flying, they had seen no land, and they were down to just a half hour's fuel remaining. Then, suddenly, they saw mountains ahead in the distance, their snow-flecked summits shimmering in the sun, the coastline dotted with islands. It had to be northern Spitsbergen.

Amundsen nudged Riiser-Larsen, who nodded. But the pilot yelled that the stabilizing controls were jamming, and they needed to land now. Amundsen spotted open water, but it was rough and wind chopped. Riiser-Larsen dropped *N25* down and over a jumble of ice hummocks, then splashed its belly onto the rolling waves of a strait between North East Land and Spitsbergen. The men whooped and cheered as they taxied through rough seas, finally arriving at the desolate shoreline inhabited only by walruses, seals, and polar bears. They tied *N25* to grounded ice and went ashore. They were all relieved to be alive and safe, but Amundsen reckoned they were in Hinlopen Strait, perhaps one hundred miles west, over high mountains and glaciers, from Kings Bay. It would be a dangerous ordeal to trek overland to get there. They ate some of their remaining food, sitting on a flat rock among boulders and washed-up driftwood.

As they rested and contemplated their recent deliverance and new predicament, someone yelled out, "A boat!" Just beyond the point of their cove appeared the mast and sail of a ship. Everyone leapt up, yelling and waving their arms, but the ship kept going. Amundsen and Riiser-Larsen told everyone to hurry back into the airplane, and Riiser-Larsen charged into the strait, bashing through waves in pursuit of the vessel. After a few rough minutes, they pulled alongside a sealer. They tied up to the ship and its shocked crew, and were brought on board by rowboat. Amundsen and his men were covered with grease, their hair and beards straggly, their Arctic parkas torn and stained with fuel oil. Amundsen reached out a blistered hand and greeted the captain. "They seemed to doubt as to who we could be," remembered Amundsen of the miraculous encounter, "But when I turned slightly round, I exposed my profile," and the captain knew instantly who it was.

"You're supposed to be dead," he exclaimed with wonder.

18

RESURRECTION

AMUNDSEN AND ELLSWORTH READ WHAT FEW PAPERS WERE available at the remote outpost at Kings Bay, where the sealer had graciously transported them. The lost Arctic aviators were front page news, with the headlines fearing the worst: "Amundsen Missing 112 Hours in Arctic" one read; "No Word of Fliers on Trip to the Pole" worried another. Now shaved, well fed, and rested, the explorers learned that rescue missions had been organized when the world had heard nothing from them after a week. Norwegian relief vessels, some loaded with seaplanes, had been cruising the open waters of north Spitsbergen. Soviet icebreakers were looking to the east. U.S. President Calvin Coolidge had assured the world he would authorize a naval plan to assist, if needed.

But as he always had, Roald Amundsen had returned from the dead. It had taken the sealing vessel a few days to tow *N25* around Spitsbergen's north coast. Strong winds had forced them to seek shelter a few times, and eventually they decided to beach *N25* and tie it down, to be retrieved by ship and transported later. Proceeding along northwest Spitsbergen, they had passed by Virgo Harbor at Danes Island, where the remnants of Camp Wellman still stood, though Wellman's grand airship hall—the *America*'s hangar—had long since blown down. Amundsen was keenly aware of the history of the place from which Wellman, and before him Swedish balloonist S.A. Andrée, had attempted flying to the pole. Amundsen had the captain hoist the ship's flags in honor as they passed, then lowered them as they sailed on, arriving back in Kings Bay after being gone, and assumed lost, for over a month. Ellsworth noticed that the hard struggle had "wrought a shocking change in Amundsen. Sleepless toil and anxiety had graven in his face lines that seemed to age him ten years."

Once news of Amundsen's return reached the world, celebration was universal. Ironically, despite having reached just shy of 88° N—one hundred fifty miles short of the North Pole, and nearly perishing on the ice—Amundsen, Ellsworth, and the others were hailed as heroes. It was the farthest north anyone had been by air, and the story of their courage, their brush with death among the floes, and their dramatic take-off amid the icebergs, touched a global chord and praise for their deeds poured in to the little telegraph station at Kings Bay. Amundsen, who less than a year before was depressed, financially ruined, and believed his exploring days were over, was now being heaped once again with laurels. He received telegrams of congratulations from the king of England, the German president, and the heads of most of the prominent geographical societies in the world. One telegram he read over and over, his chest bursting with pride. It was from his king, Haakon VII of Norway:

THE QUEEN AND I WISH YOU AND YOUR COMPANIONS WELCOME BACK. I THANK YOU FOR YOUR ENTERPRISE AND THAT YOU HAVE AGAIN BROUGHT HONOR TO NORWAY—HAAKON, R.

Amidst the flurry of praise and congratulations, Lincoln Ellsworth received sobering news that his father, James Ellsworth—who had generously financed the expedition—had died on June 2, 1925, at his villa in Florence. It was not a complete shock—his father had been declining for some time—but the news hit Ellsworth hard, particularly because his father had died without knowing his son's fate.

In Oslo, a grand celebration was being prepared for the returning explorers. After nearly a week at Kings Bay, Amundsen, Ellsworth, the four other airmen and the N25 left Spitsbergen on a coal ship. They crossed the Norwegian Sea, cruised along Norway's long western coastline, and disembarked at Horten, south of Oslo. There, the six adventurers climbed once more into the *N25* and flew up the fjord to the Oslo harbor, flanked by formations of Norwegian navy and army aircraft. Amundsen was moved to tears by what he saw. "Who can describe the feelings which rose within us as we of the *N25* flew in," he wrote, "over the flag-bedecked capital, where thousands upon thousands of people stood rejoicing." Ellsworth wept, overcome that his father had not lived to see this spectacle.

They landed and taxied slowly into the harbor filled with "the gathered naval might of Europe," including thirteen British Royal Navy ships, the warships dressed with flags, their crews hovering at the rails waving and cheering them home. Government officials met the explorers at the wharves and escorted them to horse-drawn carriages, which paraded them through city streets lined with applauding spectators. Amundsen and the others were honored later that night at a formal dinner hosted at the Royal Palace and attended by King Haakon VII and his queen.

All of the pomp and ceremony was astounding, yet Amundsen reveled in it, feeling vindication. He called that day an "undying memory . . . the best in a lifetime." Lincoln Ellsworth received a gold medal from Norway's king for saving lives by pulling two men from the polar sea. Amundsen, in a speech, praised Ellsworth as truly worthy of the medal:

> When Lincoln Ellsworth saved Dietrichson and Omdal from drowning, he saved the whole expedition; and I, therefore, deeply appreciate the King's act in conferring on Ellsworth, without whose generosity the expedition would never have taken place, the gold medal for the saving of life.

Ellsworth deeply appreciated the medal and Amundsen's words. By now they were more than just partners, they were bound in shared hardship and suffering, having forged a kind of brotherhood. Ellsworth was equally laudatory about Amundsen's leadership, telling reporters that without Amundsen's calmness, organization, and sense of purpose, they would all have died out on the frozen wasteland. Ellsworth praised Amundsen for facing death with great equanimity and approaching their dire situation with hope. "Things looked dark," Ellsworth noted, "but invariably in Amundsen's experience and in his indomitable spirit there was always most surely light ahead."

For Amundsen, the polar flight that nearly cost all of their lives had proved a resurrection. He sailed to the United States to begin a five-month lecture tour to cash in on the public favor that had swung back his way. He and Ellsworth also immediately began collaborating on a book—*Our Polar Flight*—to be published before year's end, which would chronicle their tribulations and escape from the "vast tracks of this mighty unknown whiteness." During this period, with Ellsworth also back in America settling his affairs, the two men began conspiring on another attempt to fly to the North Pole. With the death of his father, Lincoln Ellsworth had inherited a fortune, and his time spent with Amundsen—whom he admitted to worshiping as one of the heroes of his life—had only piqued his thirst for more Arctic adventure with him. He would follow Amundsen to the end of the Earth (quite literally), and now Ellsworth had his own means to back any expedition that suited him.

In press interviews, articles, and during lectures, Amundsen was now referring to his recent flight with Ellsworth as primarily exploratory in nature, essentially a reconnaissance flight to test the viability of the airplane for polar exploration. He and Ellsworth had determined that airplanes were as yet too unsafe for polar exploration. Their range and payload were limited, and as they had seen for themselves, once an airplane landed on the ice, there was no guarantee that it could take off again.

The future of flight to the North Pole, both Amundsen and Ellsworth were now saying publicly, was the dirigible. Many advancements had been made since Walter Wellman's attempts in 1907 and 1909, and by now dirigibles had successfully made nonstop crossings of the Atlantic,[1] just as

---

1 In 1919, British Airship *R34*—a near copy of the Zeppelin *LZ33*—made the first nonstop aerial crossing of the Atlantic Ocean.

Wellman had predicted could be done. For a flight to the pole, Amundsen now knew, airships were safer, and had the advantage of hovering at slow speeds without having to land on the ice, which better suited them to take photographic and other observations. But still, as of now, no one had flown an airship to the North Pole. In announcements in late fall of 1925, Amundsen vowed to be the first. There remained unknown, uncharted lands at and around the North Pole, "a million square miles of the earth's surface, unexplored." Although he and Ellsworth had been scheming behind the scenes for months, they finally publicly revealed their spectacular plan.

They had purchased a superb state-of-the-art airship from an Italian dirigible designer, and in this airship, "had agreed to attack the Arctic unknown during the following year, 1926, on a truly grand stage."

19

THE FASCIST'S AIRSHIP

BEFORE EVEN LEAVING OSLO AFTER THE CELEBRATIONS OF THEIR glorious and unexpected return from the dead, Amundsen and Ellsworth had been planning their airship voyage to the North Pole. Prior to departing for the United States, Amundsen had corresponded with the Italian engineer and airship designer Colonel Umberto Nobile, whose semi-rigid dirigible, the *N1,* was for sale. Amundsen had been intrigued by this prospect, having taken a test flight in the very craft two years prior while traveling through Italy. He had been impressed, and believed, with a few modifications, that it was perfectly suited for an Arctic voyage.

When Amundsen learned that he might indeed purchase the *N1,* he immediately sent a telegram inviting Nobile to "an important, secret conference" at his beautiful home, Uranienborg, a refurbished cottage on a quiet cove on the Bunne Fjord south of Oslo. There, along with Ellsworth and Riiser-Larsen, who had also agreed to participate in the expedition, they met Colonel Nobile, who was at the time a senior officer in the Military Air Service of Fascist leader Benito Mussolini.

Nobile was refined in dress and appearance, with a slight build and deep, penetrating dark eyes, his black hair slicked back and neatly combed. Thirteen years younger than Amundsen, Nobile was not an explorer but an academic, a scientist, and a university professor. Amundsen cared little about that. Nobile had studied electrical and industrial engineering at the University of Naples (graduating cum laude) and was now widely respected in the field of aeronautical engineering. Most important to Amundsen, Nobile had been pioneering fast and maneuverable semi-rigid airships—airships that maintained their shape partly through gas pressure, and partly by a rigid keel which stiffened the envelope. Built at Mussolini's military airship facility, the 350-foot-long *N1,* according to Nobile, could cruise at about fifty miles per hour. The airship was nearly twice the size of Walter Wellman's *America.*

At the clandestine meeting—which Amundsen and Ellsworth wanted kept private so that they might later make a bold announcement, but also so as not to alert any potential rivals of their plans and time frame— Nobile told Amundsen, Ellsworth, and Riiser-Larsen that he had been authorized by Mussolini to donate the *N1* to Amundsen and the Aero Club of Norway, provided the airship expedition would fly under the Italian flag. Amundsen flatly rejected this proposal, saying that under no circumstances would he allow this "dream of seventeen years to be fulfilled under any other flag than that of my native land." Amundsen stressed that he wished to purchase the airship "free of all conditions," and after some discussion, settled on a price of $75,000, provided that Nobile made certain modifications. The airship's nose would need strengthening and should include a special ring that would allow it to be tethered to a mooring mast. And the ship must be lightened to increase fuel capacity for a sixty-to-seventy-hour flight. *N1* had originally been built to accommodate comfortable travel over the Mediterranean for twenty to twenty-five passengers, and it possessed a luxurious interior, including a kitchen, a saloon, and an elegant bedroom. These accoutrements were too heavy and unnecessary for Amundsen's Arctic exploration needs. Nobile agreed to the changes, and the discussion turned to other important details, including projected departure dates, the makeup of the crew, and command of the expedition, such as job titles and responsibilities.

Nobile, as an agent of Mussolini's Fascist party, obviously wanted to have numerous Italians represented as crew, and Amundsen agreed to six, including Nobile. It made sense since Nobile's engineers and mechanics were trained to operate the aircraft. Then came the important issue of command. Amundsen stated that he and Ellsworth would share overall command as "expedition leaders"; Nobile would be pilot and "airship commander." This distinction was crucial to Amundsen. The orders for the handling, operation, and steering of the airship would come from Nobile, as they should. But the direction and destination of the airship must be up to Amundsen and Ellsworth. "Our function was command of the expedition," Amundsen said plainly, "and our orders as to where the ship should go must be unquestioningly accepted."

Nobile was silent, brooding for a few moments.

"Then," he said almost pleading, "will you consult my opinion?"

"Certainly," answered Amundsen, "We should be very stupid if we did not get the skipper's expert opinion of what he could do with his airship before making any decision."

Amundsen reminded Nobile that the North Pole was only part of the goal. "The whole object of the expedition," he said, "was to make the continent-to-continent flight across the Arctic Ocean—and that nothing short of obvious inability to complete that flight would [persuade] us to turn back at any point along the course." Amundsen was privately worried that Nobile might make it to the North Pole and decide, because of rough winds or weather, to turn around and return to Spitsbergen, which would be unacceptable. The main goal was to fly over the unexplored lands between the North Pole and Alaska. In the end, Amundsen and Ellsworth convinced Nobile to agree that any major decisions would be put to a vote between the three of them. Nobile knew that by doing so he was relinquishing control—as they were likely to vote together—but they would have it no other way. And Amundsen stipulated one more thing: the name of the airship must be changed from *N1* to *Norge* (meaning Norway), and the airship must fly the Norwegian flag. On this Amundsen would not budge; there would be no discussion.

After a contentious day and a half at Uranienborg, they had an agreement—with written contracts forthcoming—and a tentative departure date of March 1926 from Rome. But despite the handshake deal, clear divisions and fractiousness had surfaced already. The haughty Nobile felt

slighted to be considered as nothing more than a hired dirigible pilot, notwithstanding the obvious fact that he had designed and built the complex
flying machine. He viewed Amundsen and Ellsworth essentially as paying
passengers bent on garnering as much glory for the historic flight as possible. It was, from the start, a necessary but uneasy alliance.

Nobile returned to Rome immediately to begin modifying the dirigible. Riiser-Larsen was soon to follow him to learn airship operation and
assist with the needed modifications, as well as to hire a Norwegian crew
to build an elaborate hangar and mooring mast at Kings Bay. Ellsworth
left for the United States to manage his affairs, and Amundsen remained
home until late summer, working on the book he was coauthoring with
Ellsworth and preparing for his U.S. lecture tour. Then in early fall,
Amundsen went by steamer to America.

By March 29, 1926, all three men had gathered at Rome's Ciampino airfield
for the official transfer of the airship from the Italian government to the
Norwegian Aero Club. Benito Mussolini had organized a grand ceremony,
"a national event, with protocol and speeches." The day was declared a
national holiday, and the event was formal, with much pomp and military
grandeur. Amundsen and Ellsworth, dressed in fine tailored suits and ties,
stood next to Umberto Nobile, attired in his smart colonel's uniform. At
his side, as usual, was leashed his little terrier Titina, yapping excitedly.
Mussolini, dressed in a black suit, black tie, and bowler hat and flanked
by high-ranking military officers, headed the proceedings, giving a key-
note speech. Then came the main event: to a drum roll, the Italian flag
was removed from the keel of the streamlined, silver 350-foot airship and
replaced with the Norwegian flag. Mussolini took the folded Italian flag
and handed it to Nobile, saying, "This is to be dropped on the ice of the
Pole; prepare a fine casket to keep it in, on board." Nobile bowed his head
in deference and nodded agreement.

During the proceedings, Amundsen and Ellsworth had noticed with
displeasure that, although *NORGE* was painted in large black letters on
the airship's side, the original name, *N1,* remained in conspicuous bold

black letters near the nose and tail, and the Italian colors were still painted on the fuselage. Amundsen was miffed, but there was nothing to be done, as the *Norge* would head north in a few days on a multi-leg publicity flight with stopovers in Pulham, England, Oslo, Leningrad (now St. Petersburg), and finally across the Barents Sea to Kings Bay, Spitsbergen. Nobile and the rest of the crew would make the four-thousand-mile, one-hundred-hour shakedown flight; Amundsen and Ellsworth would travel by train and then ship, needing to attend to a few business matters as well as to make some public appearances of their own before continuing on to Kings Bay.

As Amundsen and Ellsworth made their way north, there was one other thing that rankled them. At the last minute, right before the official airship handing over ceremony, Nobile had insisted that the expedition title include his name. Though Nobile did not state it directly, it was clear that Mussolini wanted to benefit from the worldwide publicity that always attended Amundsen's expeditions, and having an Italian name attached would assure that added cache. Amundsen was initially hesitant. For one thing, he was no admirer of Mussolini and his ultrapatriotic National Fascist Party, whose thugs, the "Blackshirts," were known to physically beat up socialists and toss them from local governments. Also, Amundsen worried that shared billing would give Nobile a false impression of increased leadership, and that "the demands he made were efforts to usurp our function as commanders." In the end, not wanting to haggle over it any longer, Amundsen left the final decision to Ellsworth, who begrudgingly agreed. It was now officially the Amundsen-Ellsworth-Nobile Transpolar Flight.

When Amundsen and Ellsworth moored at the Kings Bay dock on April 21, little seemed changed since their departure ten months before. The small mining outpost and all of the surroundings were blanketed with snow, the mountains looming on all sides of Kings Bay, some up to three thousand feet high. Smoke curled from the chimneys of the two dozen brightly painted houses dotting the shoreline a few hundred yards up a sloping hill from the dock. But there was one conspicuous addition to the skyline: the enormous hangar a few hundred yards north of the town—built by the Aero Club of Norway. It had been a brutal winter, and although the Norwegian crew had worked tirelessly in semidarkness since late October of the previous year, the hangar was not quite finished. The roofless, three-sided structure was impressive nonetheless: 100 feet high, 115 feet wide, and 360 feet long. The building needed to be longer than a football field to accommodate the 350-foot-long *Norge* and shelter it from the same violent Arctic winds which had, three times, blown down Wellman's hangar fifty miles to the north at Danes Island.

Amundsen and Ellsworth moved into a comfortable little red-and-white house formerly occupied by the mining superintendent. The house was named the Amundsen Villa in honor of the famous explorer. From there, Amundsen and Ellsworth oversaw the completion of the hangar, which was trussed with sixteen steel bracing cables and encased in thick green canvas. They also saw to the raising of the nearby 115-foot-high steel mooring mast,[1] the highest structure then in the Arctic, to which the *Norge* could be tethered in its final preparations for departure. They dubbed it "The Eiffel Tower of the Arctic." At night, in the heated little house, Amundsen and Ellsworth studied the maps and aeronautical charts outlining their proposed route, the estimated flight time—approximately fifty to sixty hours depending on winds—and organized equipment and provisions necessary for a crew of sixteen should they end up, as they had in the Dornier-Wals flying boats, marooned on the ice. In the event of a mishap, Amundsen vowed to be better prepared than he had been the previous year.

Snows fell for weeks, and Amundsen spent a few hours each day out skiing to stay fit, slinging a rifle across his shoulder for the not unlikely

---

1 The mooring mast, and Amundsen Villa, are still there at Ny-Ålesund today. The mooring mast stands alone, commemorated with a plaque honoring the airship flights that took place from there. The Amundsen Villa—which was later converted into a tavern called The North Pole Bar—has been restored and is part of the historic town, now primarily used as a living museum as well as a vibrant global climate research center. It is the northernmost permanently occupied town in the world, at just shy of 79° N. The author had the good fortune to visit there in June 2023. It's a spectacularly remote and enchanting place.

encounter with polar bears. He well remembered, on a previous overwintering in the Artic, being attacked by a female protecting her cub. The bear had driven him face-first onto the ice and raked her claws down Amundsen's back before a shipmate arrived to startle her off. He called it "one of the narrowest escapes of my life."

A week after arriving at Kings Bay, Amundsen had learned via telegram that the *Norge* had successfully reached Leningrad—and to great public interest. One day alone, ten thousand people visited the hangar there to see the spectacular airship. Nobile was instructed to wait there until Amundsen and Ellsworth confirmed that the hangar at Kings Bay was ready—and then if the weather cooperated, they should fly on to Spitsbergen. On April 29, Amundsen learned from his ground crew that they had seen smoke from the funnel of a large ship steaming up the fjord. Amundsen strapped on his skis and shuffled through new snow to investigate. Sure enough, as he crested a knoll, he stooped and peered warily as the ship's steel hull smashed through the "blue and iridescent green cakes of ice . . . by the crumbling glacier of the bay." As it drew closer and anchored off the shore ice, he could see that it was the SS *Chantier,* a thirty-five-hundred-ton U.S. military vessel.

On board were American aviators Lieutenant Commander Richard E. Byrd of the U.S. Navy, his copilot Floyd Bennett, fifty crewmen, a few journalists, and an impressive Fokker Trimotor airplane named the *Josephine Ford.* It was a U.S.-sponsored aviation expedition, and they had come to beat Roald Amundsen in a flight to the North Pole.

20

THE MOST SENSATIONAL SPORTING EVENT IN HUMAN HISTORY

AMUNDSEN STOOD ON THE BLUFF OVERLOOKING THE BAY, LEANING on his ski poles and watching as the SS *Chantier* remained offshore, poised to dock. By now many of Amundsen's ground crew had made their way to the bluff as well, anxious to see who was coming. Amundsen said nothing, his eyes devoid of expression. After a time, he turned from the bay and skied alone back to headquarters to check with the wireless operator and talk to Ellsworth. After he had left, the foreman told the others to get back to work on the hangar—the *Norge* was expected in about a week.

Amundsen's expedition support vessel, the *Heimdal,* was docked at the small wharf, which could only accommodate one large ship at a time, so the SS *Chantier* circled around, edging past the pier and anchoring three hundred yards offshore. Amundsen instructed the wireless operator to inform the captain of the *Chantier* that *Heimdal* was under boiler repair and taking on coal and water, and for now it could not be moved.

An hour later, a small contingent rowed from the *Chantier* in a lifeboat and stepped onto shore. They walked briskly toward headquarters, greeted by barking sled dogs and workers murmuring in Norwegian as they passed the machine shop. The man in front was dressed in a U.S. naval uniform, "every button of his blue overcoat carefully fastened, the bottoms of his trouser legs stuffed into sturdy buckled galoshes." Just behind him strode a young man, shrugging against the cold in his woolen pea coat. Soon Amundsen heard a knock at his office door and rose from his chart-strewn desk. He thrust out his hand. "Glad you're here safe, Commander. Welcome to Spitsbergen." A few moments later, Ellsworth entered the room, "a trace of annoyance on his face, as though he was piqued at the fact that he was no longer the only American explorer in Kings Bay." Everyone sat down for a meeting.

Amundsen and Ellsworth had known Byrd and Bennett might arrive. It was hardly a secret. Months before they had actually met Byrd in New York to discuss his plans and offer some advice about the ice conditions for takeoff at Kings Bay. But at that time, Byrd was still shoring up financing, sea transportation, and trying to acquire the right airplane, so it remained unclear whether the commander would actually gather the resources in time. In the end, he had managed to garner backing from Edsel Ford (son of Henry Ford) of the Ford Motor Company by promising to name the airplane after Edsel's daughter, and from John D. Rockefeller, Jr., Byrd had pulled it off, and here they were. (Byrd was in fact still short about thirty thousand dollars, a sum he hoped to recoup with exclusive story rights he had previously sold, but he certainly did not mention this to Amundsen).

The press had made a lot of noise about this potential "Race to the Top of the World." The Amundsen-Ellsworth-Nobile Transpolar Flight had been widely publicized, with the flight of the *Norge* from Rome to Leningrad already getting daily worldwide attention. But newspapers were now also reporting on the attempt by Byrd and Bennett to succeed by ski plane where Amundsen and Ellsworth had failed the previous year in their flying boats, and journalists were hyping up a separate attempt from the other direction by Australian George Hubert Wilkins, who planned to fly from Point Barrow, Alaska, over the North Pole to Spitsbergen. There was even a French expedition planning to cross the Arctic Ocean in an innovative motor-driven sledge, an updated version of the "gasoline dogs" Walter Wellman had brought to Danes Island and tested twenty years earlier.

The Sunday March 7, 1926, *New York Times* magazine section headlined the race and called it "The greatest polar drive in history." The major story impressed upon readers what the expedition voyagers already knew: "Explorers hope to find a new continent and to open an air route between Europe and Asia."

For his part, Roald Amundsen downplayed any competitive element or rivalry between various expeditions, but that did not keep one journalist from calling it "The most sensational sporting event in human history." Neither Amundsen nor Ellsworth thought of their undertaking as a "sporting event." Attempting to fly to the North Pole and beyond, by whatever means, was a serious, life-threatening mission, and certainly no game. Still, all of the divided media attention, and Commander Byrd's sudden arrival, had Ellsworth irritated. "We had reason to be disgruntled," Ellsworth admitted later, "Byrd's flight divided the publicity from Spitsbergen . . . The *Norge* flight was costing a fortune. We needed to cash in every penny we could get as its result. We wanted complete media attention of the public . . . so that afterwards it would buy our book, see our pictures, and attend our lectures." Although Ellsworth was a dedicated explorer, he was also a savvy entrepreneur, and he projected that the total cost of the expedition he was substantially financing would run close to five hundred thousand dollars. He needed every column inch he could get.

As Ellsworth sat privately stewing, Amundsen offered to help Byrd in any way he could. The *Heimdal* would have to remain where it was for a few days, so docking there now was not possible. But with this recent storm, the ice was moving in and firming up in the bay right at the shoreline, and it would be feasible for Byrd to raft a bunch of lifeboats together and bring their men, airplanes, and equipment ashore on a kind of makeshift dock. Byrd thought that, using planking as a platform, it would probably work. But he remained concerned about taking off in the bay, which still had too much open water for his liking. "How will I get my plane in the air?" he asked Amundsen.

Amundsen's response surprised everyone. "There's a flat area in front of this house," he said, motioning with his long, grizzled hand, "You can

tramp it level and use it for your take-off strip." Byrd looked up in disbelief, then smiled suspiciously.

"You are being very generous to a rival," he said.

"But we are not competitors," Amundsen assured Byrd. "We are collab-
orators on a joint assault on the polar regions, an attack by two vehicles, one
lighter and one heavier than air. We are partners in this venture together."
Then Amundsen stood, his body language declaring the meeting over. He
had his own polar flight to prepare for.

As Byrd left to begin the difficult task of unloading and assembling the
nearly four-ton *Josephine Ford,* he grinned. Despite what Amundsen said,
in his own mind, this most certainly was a competition, and the *Norge*
wasn't even in Spitsbergen yet. Just six weeks earlier, in an article Byrd had
written for *The New York Times,* he had stated unequivocally: "All three of
us—Amundsen, Wilkins, and myself—are seeking to discover new land
and also to conquer the Arctic from the air. It is not exactly a race, but
the element of competition is there." Apparently, Amundsen had been too
busy to read the newspapers.

Early in the morning of May 7, 1926, the colossal *Norge* droned into the
upper arm of Kings Bay, its leviathan, cylindrical profile casting a moving, whale-like shadow onto the snow and ice below. Nobile was easing
the airship in slowly, "sedately as a liner" on just two engines, the third
having sheared a shaft during the last leg. Amundsen, Ellsworth, and the
ground crew had known it was coming, so they hurried to the hangar—
which had been decked out in the flags of Norway, the United States,
and Italy—to bring it in. They were amazed by its magnitude as it drew
closer, then hovered overhead, engulfing them in its shadow. Shackles were
engaged, and ropes tossed out, caught by the ground crew. A sudden gust
buffeted the *Norge,* picking some of the men at the haul ropes a few feet
off the ground, but then the airship settled on the snow-covered hangar
floor and was tied down and held steady. Umberto Nobile stepped from
the gondola and Amundsen and Ellsworth greeted him, then led him to
Amundsen Villa to debrief. Bernt Balchen, a young Norwegian Amundsen

trusted, who had been at Kings Bay the previous year helping out as well, followed along to the meeting.

Although they had yet to even start for the North Pole, what Nobile and his airship had already accomplished was impressive and unprecedented. They had completed a four-thousand-mile journey, through fog and snow and sleet and rain, the last five hundred over the wind-tossed Barents Sea, in 103 flight hours. Nobile's dark eyes darted and shifted. Fatigued and sleep deprived, he spoke quickly in Italian before remembering it was no longer his Italian airship operators and mechanics he was addressing, but a Norwegian and an American, and he switched to English, telling them about the broken crankshaft on the port engine, and some damage to the lower keel, and a torn rudder. But none of that mattered, he said. All could be quickly repaired. Amundsen sensed his anxiety and asked him what the trouble was.

Nobile caught his breath and sighed. He had spotted Byrd, Bennett, and the Fokker below as he had flown over Kings Bay. It appeared they were ready to depart on the landing strip beyond the house. The *Norge* must leave as soon as possible, he advised. Nobile clearly felt pressure, fueled by the looming specter of Benito Mussolini's expectations, to be first, to win what the world press and voracious readers were calling a race.

Amundsen shook his head. "We will not be rushed," he said, trying to calm Nobile down. "We will take every precaution . . . and leave only when the ship and weather are right."

Nobile threw up his hands in frustration, the gold braids on his uniform bouncing with his histrionics. He insisted they depart immediately. According to Bernt Balchen, who witnessed the meeting, both Ellsworth and Riiser-Larsen, who had just arrived, agreed with Nobile and "added their pleas to be first in the race." Outside the house, they could hear the loud rumblings of the Fokker's three 200-horsepower Wright Whirlwind engines as Byrd gunned the *Josephine Ford* up for its final test flight.

But Amundsen remained calm. "Our flight is not a race," he said firmly, looking past and through the men assembled as if into the skies beyond the pole. "Its purpose is much bigger than that. We're trying to chart a shorter route to the New World, and the North Pole is just a point we shall cross on the way." If Nobile had remembered their agreement to vote on major decisions, he might have taken the opportunity now, since it appeared that Ellsworth was on his side. But Nobile did not press the issue any further, and instead settled into his lodgings a few hundred yards from the hangar to get some much needed rest, agreeing to start work on the *Norge* repairs later in the day.

As each expedition worked to ready their aircraft and gear for their respective flights, the little mining settlement at Kings Bay resembled a western frontier boomtown or gold rush outpost near the top of the world. There were typically about two hundred to two hundred fifty full-time residents involved in the coal mining operation, but with the crews, staff, and participants of both expeditions, plus several journalists and motion-picture camera operators, the population in early May of 1926 had swelled to nearly four hundred. At that time, Kings Bay had a small schoolhouse, an infirmary, a laundry house, a post office, a stable, various workshops, a community building, and a telegraph station, and all were at full capacity during the weeks that the international visitors occupied the town. It was the most activity the settlement at 78° 56′ N had ever known.

The journalists and cameramen had a difficult job, mostly because the two expeditions had sold exclusive rights for stories, still images, and motion pictures to rival papers, and the competing syndicates had agreed, at least in principle, "not to take any unauthorized pictures or try to scoop each other." But in such a small place, with two main expeditions taking place—the hive of workers at Amundsen's airship hangar readying the *Norge,* and Byrd's men boot-packing the runway for the *Josephine Ford*'s test flights—the reporters constantly encroached on one another and violated their agreement in their efforts to get daily stories and then dispatch them at the single telegraph office. There were only so many good stories to be had, and the reporters were often reduced to fluff pieces ranging from weather reports to the *Josephine Ford*'s broken landing skis to the behaviors and moods of the so-called rivals, Amundsen and Byrd.

Still, the rival reporters skulked around like modern-day paparazzi in what one observer called "a kind of guerrilla warfare, with undercover operations and secret infiltrations." The reporters on each side went to great lengths, even using disguises to gain access to the opposing camps: reporters representing Amundsen donned American sailor hats to pose as members of Byrd's U.S. crew; those covering Byrd stole woolen Norwegian ski caps pretending to be part of Amundsen's hangar crew. These machinations soon became comical, especially when the presumably stealthy reporters all ended up at the only telegraph office at the same time and had to wait in line next to one another to post their nearly identical stories, often unintentionally revealing their ruses.

The behavior of the men in charge—the actual so-called "rivals"—was much more civilized. Amundsen and Ellsworth invited Byrd and Bennett to dine with them at the large officers' mess hall, and Byrd soon returned the favor, inviting Amundsen, Ellsworth, Nobile, and Riiser-Larsen to dinner aboard the *Chantier*. Despite Nobile's expressed concern about Byrd winning the race, the Italian commander found him to be "Very friendly . . . a pleasant, intelligent young man."

The wizened and notoriously gruff Amundsen could not have been more helpful than he was to Byrd's efforts. When one of the *Josephine Ford*'s skis shattered on a test flight landing, Amundsen enlisted a few of his carpenters to fashion landing skis out of hardwood lifeboat oars, and these superior skis were mounted on the Fokker. It was Amundsen, not Byrd, who had leased the lands around Kings Bay from the mining company for his hangar and mooring mast, as well as the land he had offered Byrd to use as a runway, so the grizzled Norwegian was being very generous indeed. He counseled Byrd on the prevailing winds there and the best layout for the runway, and even offered some of his own men to help shovel and level it. Given the previous year's ordeal with Ellsworth, Amundsen certainly knew something about building a runway on the ice. And being intimate with a forced landing and survival on the pack, Amundsen provided Byrd with a sledge, snowshoes, skis, and straw-stuffed winter mukluks (a trick he had learned from the Inuit) that would save their feet from frostbite, and potentially their lives, in an emergency.

Amundsen did harbor one significant worry about Byrd getting off first, one he had mentioned only to Ellsworth. He knew that if the Fokker got into trouble, and failed to return in a day or so, he would be bound by duty and honor as a fellow explorer to fly the *Norge* on a rescue mission, which would delay their own departure, perhaps until the following year. They both agreed that May offered the golden weather window in which to make the flight. "After June 1," Ellsworth agreed, "fog blankets the whole Arctic basin for the rest of the summer; and fog, by preventing us from seeing the regions over which we flew, would nullify most of the scientific value of the voyage." Waiting a year would complicate things considerably, in terms of financing and support. There was no telling whether the expedition would have to be canceled completely.

Adding additional drama and tension to everyone at Kings Bay (except, it seemed, Amundsen), on May 7 word came from the telegraph office that George Hubert Wilkins was seen at Point Barrow with his airplanes, and he wished to know if the weather in Spitsbergen was suitable for his proposed flight from Alaska, implying that he was ready to leave at any moment.

But none of these concerns caused Amundsen to rush. He was determined to wait for the right wind and weather, checking daily reports in Spitsbergen, from Russia's Novaya Zemlya and Siberia, and those coming out of Alaska, which were all equally important. While Amundsen waited, he kept up his daily fitness regimen, skiing five to ten miles a day. He also enlisted some of his Norwegian expert skiers to teach the Italian crewmembers how to ski, which might save their lives should the *Norge* unexpectedly go down somewhere out on the ice. Nobile continued fine-tuning the *Norge,* replacing the port engine crankshaft, refurbishing the stabilizing surfaces, and preparing an "anti-freeze mixture for the engine cooling system." Additionally, Nobile consulted with Amundsen and the Norwegian contingent, who prepared and packaged all the provisions and gear necessary to sustain a crew of sixteen on the ice for months if necessary.

Just after midnight, in the still, freezing hours of May 9, the cough of *Josephine Ford*'s three engines reverberated through Kings Bay. Men rose from their beds or bunks and came out to see the handsome navy-blue-and-white airplane, fifty feet long with an impressive single wing spanning sixty-three feet, taxi up the thousand-yard-long ice runway. Amundsen and Ellsworth stood on the steps of the cottage to watch as Byrd and Bennett charged the airplane forward and nosed it upward into the crystalline sky in the bright midnight sun. "She passed inland," wrote Ellsworth of the sight, "gained requisite altitude, and disappeared over the north wall of the fjord. The sound of her engines quickly faded away."

21

BEYOND HORIZONS

AMUNDSEN AND ELLSWORTH HAD JUST BEGUN EATING DINNER IN the mess hall when they heard the distinctive rumble of airplane engines somewhere to the north. "We leaped from our chairs and left our unfinished dinner," Amundsen remembered of the moment, and they sprinted through deep snow up to the runway with a few of the Norwegians who had been dining with them. None of Byrd's crew were there to watch the *Josephine Ford* glide gracefully down through the sky and skid lightly across the ice; they remained aboard the *Chantier*—also at dinner—and had not heard the airplane's final approach, which was two hours earlier than expected. So, it was Amundsen who was first to congratulate the American pilots as they hopped down from the cockpit, with Ellsworth right beside him. Byrd had made the North Pole, and history.

Ironically, given all the earlier subterfuge and machinations between the press corps, Amundsen's cameramen and photographers were the ones to actually document the event. "The only pictures of Byrd's triumphant safe return were the ones we took," Amundsen later crowed, clearly amused that Byrd's cameramen had missed the historic landing. But in the moment, he did not gloat; he embraced Byrd and Bennett, kissing them on the cheeks and exclaiming how happy and proud he was. He and his men gave Byrd and Bennett "Nine good Norwegian cheers," and all shook hands, posing for the cameras, Byrd and Bennett bundled in their

Arctic flying garb and Amundsen and Ellsworth in just the light jackets they had grabbed as they had run out the door.

Soon, everyone in Kings Bay had heard the news, and most came out to greet the returning aviators. With a good deal of sheepishness and embarrassment, the American journalists later asked for a repeat of the congratulatory greeting they had missed, and everyone posed again. Having attempted to do the same thing the year before and failed, Amundsen praised the achievement publicly, saying "Commander Byrd's exploit is one of the most remarkable on record." The two flyers had been gone fifteen and a half hours and, if their records and observations were accurate, had traveled to the North Pole and back, a distance of 1,530 miles.[1] Their flight had not been free of drama. On the outward trip one of the engines sprung a serious oil leak, and during the return flight Byrd's sextant had fallen to the cockpit floor and shattered, forcing him to navigate his way back by dead reckoning with only a sun compass.

Amundsen never doubted nor questioned Byrd's claim; that wasn't his style. He had accepted Peary's polar claim earlier, so why doubt Byrd? And besides, since Amundsen believed the North Pole already achieved, what did it matter? He had an even bigger goal—a transpolar crossing—and needed to prepare for it. Weather reports were favorable, and it was time to make final preparations for the flight of the *Norge*. Others, including Umberto Nobile, were less believing of Byrd's claim, and almost immediately after Byrd's return, though he was widely heralded as the first man to fly to the North Pole, doubts and rumors circulated around Kings Bay that his speeds and distances—given the winds, the airplane's weight, and the drag induced by the skis and floats—appeared exagger-

---

1 Like the Cook-Peary Polar Controversy, whether Byrd and Bennett actually made it to the North Pole was heavily debated at the time and in the subsequent decades. Two interesting books on the subject, for those wishing to pursue the controversy further, are Richard Montague's *Oceans, Poles, and Airmen: The First Flights over Wide Waters and Desolate Ice* (Random House, 1971); and Shelton Bart's *Race to the Top of the World: Richard Byrd and the First Flight to the North Pole* (Regnery History, 2013). Also pertinent is the following article by G.H. Liljequist of the University of Uppsala: "Did the *Josephine Ford* Reach the North Pole?" *Interavia*, Vol.15, n5 (19960), p. 589, 1960.

ated. Many concluded that Byrd's flight to the North Pole and back to Kings Bay in fifteen and a half hours was simply not possible.

Early on the morning of May 11, less than two full days after Byrd and Bennett's return, Umberto Nobile was at the hangar, anxiously shouting final instructions to his Italian flight crew. He directed his chief engine mechanic "to pour the anti-freeze mixture of water and glycerin into the radiators, and to test the engines." Three times during the night Nobile had to valve off hydrogen to accommodate envelope expansion with the rising temperatures. Nobile was frayed, his eyes bloodshot; he had been up all night readying the *Norge*. He had hoped to leave hours earlier, but winds had been too strong—"blowing diagonally across the mouth of the hangar"—and he determined not to risk damaging the airship's cruciform tail fins or envelope as it was backed out of the hangar. Exhausted, he had crawled into the control cabin and fallen asleep on the floor wrapped in a rug. When he awakened after a short nap, he sent word to Amundsen and Ellsworth that they should depart as soon as the winds died down.

The international flight crew arrived just after 7:00 A.M. There were sixteen in all—eight Norwegians (including Amundsen), six Italians (including Nobile), one Swede, and the lone American, Ellsworth. Amundsen and Ellsworth, co-leaders of the expedition, had no responsibilities in flying the dirigible. Their role, as Amundsen put it, was that of explorers, "watching the terrain below, studying its geographical character, and especially keeping an alert eye out for any signs of a possible Arctic continent." Nobile, as airship commander, managed the team of engine mechanics: navigators for steering; helmsmen running the elevators; and riggers checking the valves, ballonets, and the gas envelope in the complex operation of the twenty-ton *Norge*. It took a dozen skilled men to control and maneuver the 350-foot machine. Mechanics rode in enclosures to maintain the three 250-horsepower engines capable of propelling the airship at a proper cruising speed of fifty to sixty miles per hour for the proposed 2,500-mile flight.

Among the sixteen men on board were two radio experts running the Marconi wireless, which theoretically had a one-thousand-mile range. Amundsen planned to send progress reports as conditions allowed, though it was uncertain, once near the North Pole and beyond, whether they would be able to send or receive any messages. A Norwegian journalist was also on board to report on the flight, which, if all went well, should take fifty to sixty hours. The last passenger—though certainly not least in Nobile's mind—was his tiny twelve-pound terrier Titina, who always flew with the commander for good luck.

Nobile, now speaking in English—the one language shared, to a degree, among the crew—directed everyone to climb aboard. Amundsen was the last to arrive. He had been at the manager's office reading weather reports from Alaska. Australian George Hubert Wilkins, who had hoped to fly from Point Barrow to Spitsbergen, remained grounded there by a damaged airplane and foul weather. He reported current "violent sleet storms and turbulence along the entire coast," though the forecast called for some clearing. It would be at least two days before they arrived there, and Amundsen agreed with Nobile that they should leave, and deal with the Alaskan weather if and when they got there.

Norwegian Bernt Balchen, one of Amundsen's ground crew who had been helping load the last of the emergency gear—snowshoes and skis and food—watched Amundsen carefully as he arrived at the *Norge,* seeming to understand the import of the moment. Would this be the White Eagle of Norway's final polar exploration? Would Amundsen find a great land mass, this rumored place called Crocker Land that Robert Peary claimed to have spotted back in 1906? Balchen recorded his thoughts as his hero and countryman climbed into the gondola:

> Amundsen is the last to reach the *Norge* . . . with a roll of charts under his arm, like a scientist quietly entering his laboratory. He wears a light windbreaker, a blue flying helmet, and his familiar canvas-topped mukluks, stuffed with a fine Lapland hay called sennegrass. He pauses at the foot of the steps, and the meteorologist hands him a final weather reading.

He studies it and lifts his carved face for a moment to the sky, weighing his decision. It is a face that lived a thousand years ago, and will live a thousand years from now.

When Amundsen had taken his place in a chair by the large window spanning the front of the control gondola, Nobile gave the order to guide the *Norge,* at last, backward from the hangar. Once aloft, Nobile hovered for a few minutes one hundred feet above a snowfield near the shore, checking weight and balance, then ordered the engines started and the tethers released. Everyone in Kings Bay stared in awe at its whiteish underbelly and silvery sides suspended above the shimmering Arctic waters and glittering glaciers. They were witnessing a true engineering marvel: the pressurized envelope was taut and sleek, reinforced with metal frames at the nose and tail, with a flexible tubular metal keel connecting the two.

As the airship rose, the ground crew waved and cheered. Nobile called for more speed and the *Norge* climbed to 1,200 feet, staying over Kings Bay to avoid the high mountains on all sides. Auks and gulls flew from the high cliff walls, startled from their rookeries by the strange flying machine. As Nobile steered the dirigible toward the open northwest mouth of the bay at fifty miles per hour, Amundsen and Ellsworth looked out to see Byrd and Bennett flying alongside them in the *Josephine Ford,* then racing past, escorting them for an hour or so until they reached the polar ice pack. Finally, dipping their wings in a farewell salute, they sped back toward Kings Bay.

As the *Norge* cruised north, everyone aboard was busy. Amundsen, seated next to oval-framed pictures he had hung of King Haakon and Queen Maud, took notes in his expedition diary, marking the time, the temperature, and the geographical features below. Ellsworth observed and took notes too, sometimes getting up from his folding camp stool to stretch his legs and converse with Riiser-Larsen, who was busy doing calculations at a table with his maps, sextant, and other navigational instruments. Nobile's station was portside of the control car, where he monitored all aspects of the airship. "Behind him," wrote Ellsworth, "was a control board with signal wires to the engines and cords running to the hydrogen exhaust valves and ballonets." He periodically left his post to consult with the engine mechanics and riggers.

The mechanics stayed in their housing compartments with the engines at the rear of the *Norge,* while the riggers moved about the internal network of gangways and catwalks checking the ballonets and valves inside the massive envelope. One rigger's job was particularly daunting: he had to periodically inspect the gas valves at the very top of the airship, and to report on any ice forming outside the envelope. Nobile commented on the unenviable, dangerous task: "He had to go through a small window in the bow of the ship, climb up a steep steel ladder on the outside, and, in an icy wind blowing nearly fifty miles an hour, crawl along the top of the ship, from bow to stern, for a length of 70 to 80 yards, clinging with one hand to the guide rope that had been fixed there."

As they navigated along Spitsbergen's northern coast, the wireless operators sent messages of their location and progress, and the Norwegian journalist filed a few reports which were sent out. Twenty years earlier, Walter Wellman had been the first to send short wireless messages from the Arctic, but with the improved technology and range, the journalist aboard the *Norge* was able to dispatch entire stories, published in newspapers while the airship was still in flight.

The stories ran in newspapers around the world, and as might have been predicted, both the Italian press and the Norwegian press used the moment for national pride. The Italians were brazen to the point of fault. The front page headline of the May 11 issue of *Il Piccolo* read as follows: "Under an Italian flag, in the Spirit of Fascism, Norge Sails into the Polar Sky." In truth, at that exact moment it was the Norwegian flag snapping at the tail of the *Norge* in the sub-zero air. And inside the airship, the international team was flying in the spirit of cooperation and shared vision, not fascism. But the Amundsen-Ellsworth-Nobile Transpolar Flight had become a deeply political and nationalistic symbol for Italy and Norway, as well as a sensational story selling hundreds of thousands of papers across the globe.

The men aboard the *Norge* had other things on their minds. Riiser-Larsen took land and radio bearings, "then set the sun compass for the North Pole." Between cups of hot coffee to wash down meatball stew and frozen boiled eggs, Ellsworth gazed out the window at the ice pack, where he and Amundsen and the others had survived for over a month the previous year. "It was just as it had been," he wrote of seeing it again, "piled and broken, a vast sweep of white desolation, silent, mysterious." He spotted a few polar bear tracks, a few seals, a single Arctic fox, but no other life in the region that had engulfed one of their Dornier-Wal flying boats and nearly taken their lives.

As the *Norge* whirred north, Nobile dealt with a few complications. At one point the port engine quit suddenly. Nobile ran the airship on two engines until the mechanic discovered the problem: the carburetor had been clogged by ice, which he managed to break loose. Soon the engine was up and running again. Then, not long afterward, Amundsen and Ellsworth noticed ice forming on the celluloid windows, completely obscuring their vision, and Nobile's. In minutes, reported Nobile, "all the metal parts of the ship were covered in ice . . . and on the exterior of the control cabin." This was dangerous; not only did the ice make the airship too heavy, but shards could potentially fly off and tear the envelope or damage the propellers—or both. Either could prove fatal. Nobile ordered a fast ascent to three thousand feet above the fog, where the ice gradually dissipated.

At just past midnight, with the sun visible beyond the horizon due north, Ellsworth realized it was May 12—his forty-sixth birthday. Inadvertently, he yelled out the news, and the journalist heard him and had the wireless operator place a quick dispatch out to the world. Lincoln Ellsworth was certainly the first person to celebrate his birthday in a dirigible near the North Pole. A few minutes later the wireless operator handed Ellsworth a wireless message: "Congratulations! Your friends in Kings Bay."

An hour later—sixteen and a half hours out of Kings Bay—Amundsen,

Ellsworth, and Nobile all noticed Riiser-Larsen kneeling and peering intently out his window, "bent over the sextant in his hands." Nobile descended through a hole in the clouds, dropping to just 750 feet above the pack to get Riiser-Larsen a clear shot of the sun. The two other navigators were also engaged checking and rechecking their position based on the magnetic compass, prior radio bearings, and their own sextant observations. The cabin drew eerily quiet, with only low, constant humming of the engines outside. Everyone knew what was happening and no one spoke. Nobile brought the *Norge* down to six hundred feet, and then Riiser-Larsen, pressing his eye to his sextant, saw the image of the sun covering the bubble. "Here we are!" he exclaimed. Nobile descended lower still, then had the engines shut down. "Ready the flags!" he and Riiser-Larsen yelled in unison. It was 1:30 A.M. on May 12, 1926.

The silent dirigible, running at a slow idle, hovered there, just above the icy surface at 90° N, an unfixed, ever-moving place of immense significance. Amundsen and Ellsworth went to the gondola window with their flags (Ellsworth's had been given to him in a ceremony by U.S. President Calvin Coolidge). The riggers and engineers—everyone on board—came forward and removed their hats. The flags had been affixed to heavy metal vanes to ensure they stuck in the ice. Amundsen dropped the Norwegian flag, then Ellsworth tossed his American flag. Nobile appeared, having removed the large Italian flag from the casket he had promised Mussolini to protect it in. Nobile let it go, and soon all three metal stakes were impaled in the ice next to one another, their multicolored banners waving in the Arctic wind.

22

FLYING BLIND

FROM THAT SINGULAR PLACE AT THE TOP OF THE EARTH, THE GEO-
graphic North Pole, all directions led south. Nobile had the engines
brought up to half speed, and for an hour they slowly circled around
the pole, again and again. Amundsen and the navigators dutifully recorded their readings in triplicate; there would be no controversy as to
the veracity of their location. The wireless operator sent a message of their
accomplishment—the last of more than fifty that made it through to
Svalbard Radio—and soon the news reached telegraph offices and news-
rooms around the world and was being printed and broadcast by radio
from sea to sea.

The historic announcement from the North Pole made its way into
the hands of Walter Wellman, who had not been in an airship since his
last flight in *America* sixteen years before, but who had as early as 1893
predicted that "aerial navigation will solve the mystery of the North Pole
and the frozen ocean." He had also sent the first wireless communication
from the polar regions. On the same day the world learned of Amundsen's
triumph, the Arctic airship pioneer sent the following message:

New York, May 12, 1926
Roald Amundsen:
As first navigator over polar ocean ice pack in dirigible
airship, my hearty congratulations upon your

great achievement, vindicating judgment of pioneers
in that field of exploration. WALTER WELLMAN

By then, the *Norge* had lost communication with the world, but Wellman was still proud to see his prediction finally validated.

As the *Norge* passed over the fluttering flags one last time, Amundsen glanced over to his countrymen Oscar Wisting, who was working the wheel as helmsman. Wisting looked up, and the two men stared knowingly at one another, though neither said a word. Wisting had been with Amundsen at the South Pole fifteen years before. Amundsen came forward, patted Wisting on the shoulder, then shook his hand. Though the gesture was subtle, everyone in the cabin knew what it meant. Roald Amundsen and Oscar Wisting were the only two men in the world to have been to both the South Pole and the North Pole.[1]

Once Amundsen and Ellsworth had confirmed with the navigators that their sextant and compass readings, plus their times and positions, were accurate and had all been recorded in the log books, Nobile set a course for the Alaskan coast, and they flew on into the unknown regions beyond the pole. This area held the most interest for Amundsen and Ellsworth as one of the last places on the Earth's surface still unexplored and unmapped. The two men drank coffee and sat peering through binoculars from the control car window, scanning for any land masses below as the *Norge* flew smoothly through blue skies at 1,800–2,400 feet, giving them a clear view of more than fifty miles in all directions. For hours, staring and blinking, their eyes burning with fatigue, what they observed below was all ice, "nothing but a chaos of barricades and ridges," rifts formed by winds and currents erupting from the frozen plain, but still they kept

---

1 The geographic North Pole is the end of the Earth's rotation axis and is the North on geographic maps. This pole lies in the middle of the Arctic Ocean. The magnetic North Pole is the point where the lines of force of the Earth's magnetic field converge. This is the point that attracts the needle of a compass and is not the same as the geographic pole. The magnetic North Pole was located for the first time in 1831 by John Ross in the Canadian High Arctic. Ross was exploring the Northwest Passage by ship when his vessel became stuck in the ice for four years. The magnetic North Pole is continually moving.

their vigil. Their objective was either to discover new lands or disprove their existence.

At one point, after countless hours of surveying, someone called out that they could see a mountain ridge rising out of fog gathering to the west. Nobile altered course, and all gazed expectantly through the windows, straining to see it. Then Amundsen chuckled out loud, setting down his field glasses. "It's just a Cape Fly-Away!" he said, alluding to one of the numerous mirages and optical illusions occurring in the Arctic. As the imagined mountain dissolved before their eyes, Nobile corrected course and on they flew toward the south.

Opportunities for sun sightings were reduced as the fog thickened, and eventually the *Norge* was completely enveloped in a miasma of mist and vapor. Nobile noted that the airship was becoming heavy, and his crew confirmed his fears: ice was forming on all the metal parts of the dirigible, including the solar compass and the 450-foot-long radio antenna trailing behind the gondola, and they had lost all communication. "This was serious on two counts," wrote Ellsworth of the dilemma. "It stopped our weather reports . . . and it deprived us of radio bearings, leaving our navigator at the mercy of the magnetic compass and such sun observations as he might be lucky enough to make." Now, if they were driven down onto the pack, they would have no way of reporting their position. They were flying blind. And although no one aboard the *Norge* knew it at the time, once they had stopped transmitting messages, relief ships and icebreakers in Alaska were made ready as a result of the *Norge*'s silence.

Then, suddenly, everyone heard an alarming noise, a blast like an explosion. A mechanic rushed forward to inform Nobile that chunks of ice were falling off the outer shell of the envelope and striking the gondola and the propellers, and shards of ice had rent holes in the gondola and the airship's envelope underbelly beneath the keel. Riggers rushed along the interior catwalks with cementing adhesive to patch the tears, and Nobile called for slower speed on the two outer engines, hoping this would reduce the impact of the ice. Large enough ice fragments might also damage the propellers, forcing them to land on the pack surface. Nobile's greatest fear was that sharp ice fragments would "pierce the walls of the gas chamber," causing serious loss of hydrogen, or worse, an explosion.

Flying along through fog and heavy snow, they endured a constant barrage of ice striking the ship. Ellsworth listened with fear and fascination: "By standing up in the cabin," he wrote, "we could hear the pieces hit the taut fabric . . . thus hour after hour the *Norge* battled on." Periodically, the fog dispersed enough for Amundsen and Ellsworth to see below, and the entire surface, as far as they could see, remained a patchwork of crenelated ice, with occasional water leads snaking and winding through the pack, but the expanse contained no land masses. Nobile, despite his fears for the airship's imperiled condition, noted the stark beauty below: "The surface of the limitless sea of ice—all white—seems veiled in a transparent whitish mist. Here and there the whiteness is streaked with blue—that tenuous shade of blue . . . and long serpentine channels, dark and grey in color . . . like a wide black river, its banks formed by layer on layer of blue-sprinkled ice."

To combat the icing, Nobile descended very low, trying to get beneath the fog-filled air. Now Amundsen and Ellsworth could see the water leads widening, with large loose floes and eventually stretches of open water. Riiser-Larsen, who had done a masterful job navigating by the previous day's weather map to account for wind drift, and the magnetic compass for direction, managed to take a sun-shot early that morning. He reported that they were holding steady on their course to the south. Nobile remained concerned: no matter how low he flew, ice fragments continued to rip the bottom of the envelope, and his mechanics had run out of adhesive cement and could no longer make repairs.

A few hours later, Riiser-Larsen belted out, "Land ahead to starboard!" Everyone in the control car gazed below and could see mountainous terrain in the distance, long slate-gray rock ridges, and as they drew closer, smooth shoreline ice. They had been flying about forty-seven hours and had covered two thousand air miles in reaching the Alaskan coast. Amundsen and Ellsworth shared mixed feelings. Relief at not having been forced down onto the ice again, and some resignation at having discovered no new lands. "We could conservatively claim," wrote Ellsworth, "to have looked down upon a hundred thousand square miles of unexplored territory. We could tell geographers that there is no land at all between Alaska and the Pole. We had established the scientific fact that the North Polar Region is a vast, deep, ice-covered sea. The white patch on the top of the globe could now be tinted blue."

Amundsen believed that the land below "looked like Point Barrow country," where he had been before. There was some discussion of attempting to fly inland, over the mountains, but with uncertain elevations and no new weather reports, that was ruled out as too dangerous. Amundsen, Ellsworth, and Nobile all agreed that instead they should follow the coastline southwest, bearing just offshore, if possible, to avoid mountains. Nobile would try to stay low enough to view, as daylight provided, landmarks and features which might affirm their exact position, and then set a course for Nome. Winds picked up, hammering them from the north, pushing them with a fifty-mile-per-hour tailwind into furious weather. "Thereafter," wrote Ellsworth in his journal, "through a black night and a long, wild, fog-choked day we battled for our lives."

With the intense turbulence, the airship rolled and pitched and yawed, and the cabin floor became a mess of strewn food and thermoses. The inside of the control car grew cold and damp with condensation. At last, glassing out his window, Amundsen spied a cluster of houses below that he could confirm was Wainwright, Alaska. He first recognized a reindeer farm, then a house where he had lived for a time three years before. As they flew over, they could see people outside—and even up on the roofs— waving at them as they blew past. Among the Inuit people gathered there was also Australian George Hubert Wilkins, who, it turned out, had never entered the so-called "greatest sporting event in human history" due to foul weather.

As they hurtled along through freezing rain in dim light, attempting to remain low enough to maintain visual contact with the coast, elevated promontories and jutting headlands began appearing dangerously close. At one point, as they soared hazardously low, the dangling antenna, still rimed by ice, struck the landmass, and Nobile knew they must either go higher or head west, away from the coast, to avoid a collision with a mountain. The driving winds, now reaching gale force, made the decision for Nobile, pushing the *Norge* out into the Bering Strait.

After more than sixty hours in the air, all of the men were exhausted and sleep deprived, and some began hallucinating, imagining land and mountains despite being over the sea. The wind wailed through small gaps in the control house portals, creating weird sounds, "at times like a trumpet," Nobile recorded in his journal, and at times "like a factory siren." They were driven so far west over the upper Bering Strait that Nobile estimated they were now closer to the coast of Russia than Alaska, and he urged his helmsman to steer back toward the east, battling a strong cross wind and rough air that tossed them up and down, sometimes as much as two hundred feet in a matter of seconds.

Staying as low as possible over the ocean, they watched below as wild, wind-tossed whitecaps gave way to compacted ice once more, and Nobile knew their altered course had finally brought them back toward the Alaskan coastline, where they spotted frozen inland bays, and beyond those, low rocky hills that became high mountains farther inland. Careening south down the coast, sometimes through the fog and cloudbanks they saw the huts of remote Inuit villages, and the forms of people moving about, but the sightings were too fleeting and indistinct to identify an exact position. They weaved and rose and fell, at times so near hilltops that twice more the antenna struck land, and Nobile knew he must take the airship higher. Riiser-Larsen agreed that above the fog he might get a sun shot with his sextant, and they started ascending.

They broke through into sunshine at nearly four thousand feet, but the sun was too high in the sky for Riiser-Larsen to get a good observation from the control car window; the nose of the *Norge* was in the way. So Riiser-Larsen bravely scrambled into the keel and made his way forward to the hatch, where he climbed out, sextant and instruments in hand, and up the metal ladder on the outside of the airship, blasted by sub-zero wind. "He made his observation," Ellsworth recorded with admiration, "on the flat glacéed top of the envelope." Riiser-Larsen's courage paid off, and his sighting showed them at latitude 67° 5′ N. By checking their charts, this placed them just north of the Arctic Circle near a tiny village called Kivalina. Maps and coastal pilot books showed Nome still over two hundred miles south, with the Kotzebue Sound and high mountains in between.

Just as they confirmed their position, Nobile yelled out that the ship was becoming light and rising too fast. The sun was warming the hydrogen gas, expanding the envelope to near its bursting point. Nobile watched the manometer gauges rising and quickly opened the valves, but still they kept climbing, the nose pointed upward, rising to 5,400 feet despite Nobile spinning hard down on the elevator wheel. Seconds were critical. Countless airships had exploded when they passed pressure height. He yelled in Italian for everyone to move forward, desperately hoping to add weight to the nose, but for a few tense moments the Norwegians just stood there, dumbfounded, until he repeated the command in English, and all members scurried along the catwalk to the front of airship, some dragging heavy crates of pemmican tins and gasoline cans for added weight. The nose finally tilted downward, and *Norge* started to descend. Nobile watched the pressure gauges back down, and he closed his eyes and let out a gasp of relief.

His relief was short lived as the airship dove straight down and back into the fog, plummeting nearly five thousand feet in just minutes, where they could make out looming mountaintops once more. Nobile called for Riiser-Larsen to steer them hard west, back out over the sea to avoid a crash, and after a time, white-knuckled and breathless, they were again above water, flying through clearing skies. Nobile, utterly spent, slumped down into a chair and closed his eyes.

Though damaged by multiple ice impacts, the antenna still functioned periodically, and eventually they made radio contact with Nome, which informed them of calmer weather there, and said there were many volunteers willing to work as ground crew to help them land when they arrived. Eventually Amundsen recognized the Serpentine River below, "Which from the air," he noted, "is impossible to mistake, with its distinct snake-like twists."

Now they knew they were about one hundred miles from Nome. But it remained too dangerous to fly directly overland, so they navigated the longer coastline route around Cape Prince of Wales.

The exhausted crew flew the beleaguered airship through intermittent gusts, hoping to make Nome, but Amundsen was actively looking for any suitable place to land. With Nobile's consent, riggers built "a sort of sea anchor—a canvas sleeve sixty feet long filled with all sorts of heavy objects that gave it a weight of 600 pounds." To its end they affixed two ice anchors. It was no substitute for a proper mooring mast, but their options were limited and they hoped the ice anchors would grab land and hold them steady so they might hover, release hydrogen gas, and make a safe landing. They passed over bleak, gray shoreline and low hills and saw the ominous image of a three-masted ship, lying on its side, abandoned in the ice. The weather was worsening again, and the airship pitched and bucked in snow flurries and a stiffening headwind. Riiser-Larsen had been awake so long that, looking out of his window with field glasses, he told Amundsen he could see horses and uniformed cavalry men on the beach. Amundsen glassed the area and laughed at Riiser-Larsen's visions. "They were only irregular brown stripes in the sand along the coast."

Soon they were again above a scattering of houses and people. It was obviously too small to be Nome, but Amundsen, Ellsworth, and Nobile agreed it was time to get the tattered *Norge* down, as they had only seven hours of fuel remaining, and darkness was coming on. A long, flat stretch of shore ice appeared to be suitable, and Nobile called for reduced engines, and they hovered into a strong wind. Villagers ran below, waving their arms, and at five hundred feet riggers dropped landing ropes and started lowering the makeshift ice anchor. Just as it hit the ground, a wind burst tossed them shoreward, yanking the ice anchor into the air and then dragging it along on the ice, but when the gust subsided the anchor caught and held. The dozen men scrambling around below the *Norge* caught hold of the haul ropes, and Nobile yelled out to stop the engines and release hydrogen.

The mammoth airship dropped onto the ice, bouncing once on its rubberized, air-filled shock absorbers beneath the gondola, then settling down slowly as Nobile valved off more gas and the envelope deflated. The *Norge* groaned and hissed. "And the great bag," recalled Ellsworth, "its ice-sheathed fabric sagging into its ribs, came to rest, a hundred yards from the nearest cabin." One by one, Amundsen, Ellsworth, and the crew climbed down from the gondola and staggered onto the shore ice. Nobile was last off, with Titina yipping and racing around on the ice. The men stood next to the sagging airship, which had borne them from Spitsbergen to the North Pole and across the Polar Sea—2,700 miles—in seventy-one hours. Their eyes were crusty and bloodshot, and they were hungry, having eaten nothing but chocolate bars and chunks of cold pemmican for three days.

Amundsen approached one of the locals looking on with awe, stunned by the bizarre sight of the stupendous airship, slumped over onto its port side "like the skeleton of a beached whale."

"Where in the world are we?" Amundsen asked the man.

"Teller," he replied.

With that, Amundsen and Ellsworth turned from the *Norge* and stumbled, stooped and stiff-legged, toward the village.

23

BAD BLOOD

AMUNDSEN AND ELLSWORTH HAD LEFT NOBILE ON THE BEACH SU-
pervising the dismantling of the badly damaged *Norge*, while they hurried
to find the telegraph station. They wanted to contact *The New York Times*,
with whom they had an exclusive contract, and get the first word of their
successful flight out to the world. And indeed, the world was waiting for
news. "No News from *Norge*—Missing Two Days" *The New York Times*
had exclaimed on its front page when the *Norge* stopped transmitting messages. Commander Richard Byrd and his copilot Floyd Bennett, still at
Kings Bay, had readied the *Josephine Ford* for a rescue flight. But Amundsen and Ellsworth discovered that the telegraph station was inoperable in
Teller, the small hunting, fishing, and reindeer ranching outpost of just
a couple of hundred residents. They needed to get to Nome as soon as
possible.

The expedition team leaders found lodging with local inhabitants
and spent their first night back on terra firma together. They enjoyed a
hot meal, and, according to Nobile, "Everyone was in high spirits." They
had reason to be. Although they had discovered no new lands, they had
solved a centuries-old geographical mystery, and, as Nobile put it proudly,
"Through the large black patch which indicated these regions on the Arctic charts, showing them as inaccessible, we had drawn a line of light . . .
1,200 miles long . . . We had proved that there is no continent in these
parts, but a frozen sea—the Arctic Ocean—and we had been the first to

cross that sea." The explorers all congratulated one another, and for the rest of the evening, there was a deep sense of shared accomplishment. One chronicler heralded the journey of the *Norge*—which bisected the greatest unexplored region on earth—as "on par with the important discoveries in the history of mankind."

After two days, Amundsen and Ellsworth left Teller, while Nobile remained behind with Riiser-Larsen to salvage what they could of the *Norge*. Amundsen and Ellsworth initially went by dogsled, traveling for fifteen miles over ice to reach open water, towing the motor launch *Pippin* that would transport them the rest of the way to Nome. It was strange and incongruous to see Amundsen behind a pack of sled dogs at this seminal moment. His flight on the technologically advanced *Norge* had definitively signaled the end of the Heroic Age of Polar Exploration and proved, as both he and Walter Wellman before him had predicted, that the airship had "supplanted the dog," and yet there he was, riding along fur-clad in the old-fashioned way, behind a team of frothing and snarling huskies. In a way it was fitting, the last of the Vikings with one grass-lined canvas mukluk in the old explorer's world, and the other in the new aeronautical world he had helped to usher in.

Amundsen's reception in Nome was unexpectedly tepid, mostly because he had not arrived, as he had said he would, by airship. The local authorities had anticipated being the terminus of the *Norge*'s historic "Rome to Nome" seven-thousand-mile flight and had invested time and money preparing Nome's airstrip for their arrival. A welcome archway had been built downtown, through which the returning heroes were to be promenaded, and the town had been decorated with banners, flags, and signs. But when Amundsen arrived by boat, only a few locals—some of them previous acquaintances—were there to greet the returning hero at the dock. Having been "missing" for two days, the world had been on high alert, and yet once again the intrepid Roald Amundsen—now the undisputed greatest explorer of all time—had returned from the dead. But this time, his reappearance, at least in Nome, felt subdued.

At the moment, Amundsen and Ellsworth were less concerned with their reception than with getting their story to the editors at *The New York Times*. They ensconced themselves in town, renting a room at the Nome Hospital for a writing office. Fortunately for them, they soon learned that Fredrik Ramm, the Norwegian journalist who had accompanied them on the *Norge,* had been posting initial dispatches to *The New York Times* from Teller. The *Norge* wireless operators had managed to get the wireless station there, which had not been operational for two years, working again. Ramm's first stories told of their hovering over the North Pole, their confirmed records indisputably locating them there, the planting of the flags, and their harrowing last twenty-five hours of flight. A rapt reading public devoured the dispatches, and all seemed well, until Amundsen learned what Nobile had been up to since their arrival.

Before the journalist Ramm had managed to send his stories, Nobile had sent news of his own to Rome, breaching his own written contract with Amundsen. As a result, on May 17, 1926, 100,000 cheering Italians gathered at the Piazza Colonna to hear Benito Mussolini, standing next to a large Italian flag—and smaller Norwegian and U.S. flags—praise Nobile and highlight "The excellence of the dirigible, which was conceived, designed and constructed entirely by an Italian air force . . . adding another glory to the Italian flag." In fairness to Il Duce, in a subsequent speech given to the Italian senate, Mussolini was generous in his praise for Amundsen and Ellsworth, calling them "Superior beings and heroic men," but by then Amundsen was livid; Nobile had scooped him, and had broken explicit terms in a binding contract by doing so. Nobile's contract with Amundsen and the Aero Club of Norway prohibited any "radio or other telegraphic communications sent from the airship or from land stations during stops," nor was he to send anything out prior to the official Norwegian press reports, which he had already done.

The brief camaraderie that had existed the night of their landing at Teller was over, and an ugly war of words, a full-fledged battle of personal attacks, aspersions, claims, and counterclaims would soon reach fever-pitch. From Teller, Nobile sent Ellsworth a confidential letter, by boat, complaining bitterly "that you and Amundsen are signing the *Times* articles without my signature." While that was certainly true, Nobile had no grounds for complaint. Amundsen and Ellsworth held the exclusive contract with *The Times* and were in no way bound to include his name in their bylines. In fact, the editors at *The Times* would never have attributed writing credit to Nobile since he was not the one writing the articles.

Nobile remained in Teller for nearly three weeks dismantling and packaging the engines and other usable parts of the *Norge*. When that work was finished, Nobile sent a wire to the Coast Guard at Nome asking that they come retrieve him at Teller, a request that was granted. But when the Coast Guard cutter arrived, Nobile boarded with only the Italian crew members, leaving Riiser-Larsen and the remaining Norwegians to find their own passage to Nome. Nobile had also contacted a Catholic priest in Nome and organized a "grand fête for the Italian members of the expedition" that included most of the city and a great deal more fanfare than Amundsen had received. Adding insult, Amundsen and Ellsworth had not even been invited to attend the celebration. When Nobile learned where Amundsen, Ellsworth, and the Norwegians were staying, he declared the quarters "beneath the dignity of an Italian officer," and requested finer lodgings. The best hotel in town was closed for the winter, inhabited only by the host and his family, but Nobile demanded it be opened for him and his Italian crew, where they were accommodated, as Amundsen put it, "in lonely grandeur."

During the flight itself, everyone had worked well together, and although weather and icing had created considerable tension and drama, all of their goals had been achieved. But now, in the immediate aftermath, a sharp wedge had been driven through the expedition, one that would culminate in a vicious fight for the spoils. For the remainder of their stay in Nome, Amundsen and Ellsworth spent most of their time holed up in the rented office at the Nome Hospital, writing the first draft of their book *First Crossing of the Polar Sea*. They invited contributions from Norwegians Hjalmar Riiser-Larsen and Oscar Wisting, and from Swedish meteorologist Finn Malmgren, but none from Umberto Nobile, who was now haughty, distant, and intractably incensed, feeling he and the Italians were not receiving due credit in the American press.

On June 17, the now bitterly divided crew of the Amundsen-Ellsworth-Nobile Transpolar Flight boarded the steamship *Victoria* for Seattle. As the ship sailed south into Norton Sound, Ellsworth approached Amundsen, who was standing at the rail, gazing out wistfully at the coastline of the rugged land he loved and revered. Ellsworth stood beside him and noticed that his eyes were wet with tears. The two men were quiet for a long time.

"I suppose I will never see it again," Amundsen said at last, his voice tinged with regret.

The passage through the Bering Sea and around the Aleutian Islands took twelve days because of bad weather and heavy ice. The journey was also made awkward because the three expedition leaders were often asked to pose together for photographs, and they did so out of obligation, typically with Ellsworth in the center, separating—both physically and symbolically—what were now two combatants. Then the three men would drift apart without speaking.

Nobile was not idle aboard the *Victoria*. He sent messages to the leaders of the large Italian community in Seattle, hoping to organize a strong turnout for their arrival. Amundsen worked on notes for a lecture tour he intended to embark on in the East Coast. Ellsworth spent time on deck gazing south, thinking about what he would do next. He was only in his mid-forties and still dreamed of other adventures. "I could not sit down and vegetate," he wrote of his frame of mind then. There remained more exploring for him to do.

Amundsen and Ellsworth had expected a rousing reception in Seattle. By now the story of the *Norge* flight had dominated global headlines for six weeks, and they anticipated a returning heroes' welcome in Seattle. But even they were surprised by what they saw as they stood on deck while the steamship docked. "We were not prepared," wrote Ellsworth, "for the vociferous welcome we received, the harbor was black with launches jammed to the gunwales with cheering men, women, and children." Amundsen and Ellsworth approached the rail, ready to bow and wave and acknowledge the crowd, but as they did so, no one seemed to pay any attention to them. Most of those assembled along the wharf were looking up beyond them, to the bridge, waving, and Ellsworth noticed that "the cries from below were in a foreign tongue," the bravos coming in almost ear-splitting waves. Amundsen and Ellsworth looked up to the bridge and saw Nobile there, standing upright and proud, dressed in his finely pressed military uniform, with Titina yapping at her leash. "Every time he lifted his arm in the Fascist salute," wrote Ellsworth, "the huzzas rose from below."

Amundsen and Ellsworth stood there, ignored and dumbfounded, dressed in shabby miners' garb they had purchased in Nome. Nobile had told them at Kings Bay that for weight restrictions, crew members were to bring no extra clothing beyond their Arctic flight garb, but he had secreted away his fancy uniform for just this occasion. It appeared to Amundsen premeditated, an intentional snub, made worse when, as they disembarked, the official reception committee standing on the quay "failed to notice the drab polar explorer, he who towered above all polar explorers, and handed the town's welcome bouquet to . . . an officer with a funny little dog at his feet." Nobile received the flowers, nodded in respectful thanks, and raised another Fascist salute to uproarious applause.

None of this should have mattered, but Amundsen took it personally. He had toiled in the Arctic and Antarctic for decades, had wintered over in the coldest places on Earth, had hacked an escape channel for the *Belgica* through sea ice, had been to both poles. Umberto Nobile had made one transpolar flight. Amundsen knew that optics mattered, and although he fumed inside, he did nothing overt about Nobile's showboating. "Notwithstanding my indignation and contempt," Amundsen wrote of the situation, "I made no effort to thwart him . . . it was beneath my dignity to enter a competition for the moment's precedence with this strutting upstart."

Amundsen and Ellsworth immediately purchased appropriate suits for a few awkward luncheons and public appearances they still had to make in Seattle with Nobile, but soon, the uneasy alliance parted company. Amundsen and Ellsworth, along with Riiser-Larsen and the rest of the Norwegian crew, left by train for the East, after which Amundsen and his crew would return to Norway. Amundsen and Ellsworth needed to remain together to finish drafting *The First Flight Across the Polar Sea,* which their editors in New York hoped to publish in early 1927.

Umberto Nobile had originally planned to leave Seattle for Japan to consult and train the Japanese Navy on an airship the Italian government had built and sold to them, but Benito Mussolini sent Nobile orders to remain in the United States for several months to make a lecture tour in cities with large Italian populations. The cooperation between the Italian and Norwegian crews on the *Norge* soon devolved into a dueling publicity tour between the two factions, with American Lincoln Ellsworth initially staying above the fray. Nobile, for his part, saw the schism this way:

> Norwegians and Italians left Seattle separately and were celebrated by their respective immigrant colonies as they continued eastward through the USA . . . Mussolini used the occasion to make political capital among the USA's ten million Italians. Then the inevitable happened, namely that both Italians and Norwegians emphasized their own importance in connection with the successful voyage, often by detracting from the others' contribution. In this way in time the split became deep-rooted on both sides.

In New York on the Fourth of July, 1926, Amundsen finally received the reception and adulation he deserved. He and Ellsworth and the Norwegian crew arrived at Grand Central Station where "bands and a thousand unending ovations greeted them," and they were driven through the city with a police escort, paraded through the streets to the Norwegian-American pier in Brooklyn. "A crowd of several thousand," reported *The New York Times,* greeted the explorers and cheered them. The great Norwegian explorer was smothered under armfuls of roses until he looked like a moving flower bed." Amundsen beamed broadly and waved to the crowd. He was clean-shaven and dressed smartly in a tailored double-breasted suit for the occasion.

As Amundsen and Ellsworth shook hands with members of the crowd, Amundsen spotted a man in a deep blue U.S. Navy uniform. "Byrd!" he called out. "Byrd, come here!" Richard Byrd had just recently returned from Kings Bay and had come to New York to welcome Amundsen and Ellsworth. Amundsen pulled him through the crowd and slapped the fellow explorer on the back, then he and Ellsworth posed for pictures with Byrd. Each praised the other's achievements as cameras flashed and they answered questions from the press. Byrd called the *Norge* feat "the greatest non-stop flight in the history of the world," while Amundsen referred to Byrd's flight as "Beautiful . . . A wonderful piece of navigation." The mutual admiration and fellowship was in stark contrast to the tense media sessions Amundsen had recently endured in Seattle. But the question on everyone's mind was what Amundsen's plans were. What was next?

"I'll never explore again," he stated flatly, and murmurs and whispers went round.

"Never?" a reporter shouted back. The crowd at the pier fell silent.

"Never," Amundsen confirmed.

24

"HONOR AND GLORY
ENOUGH TO GO AROUND"

FOR ROALD AMUNDSEN, THE AIRSHIP CROSSING OF THE ARCTIC
Ocean bookended the most remarkable career in polar exploration history.
He was fêted in Norway, Japan, and the United States. "Amundsen's Feat
Caps His Career" read one headline on news of the *Norge*'s safe arrival at
Teller. His life's work was fulfilled. He had achieved every expedition goal
he had set for himself since childhood, and there remained nothing left to
conquer. At fifty-four, he was officially retiring from exploration. "All the
big problems are solved," he said definitively. As to whatever smaller prizes
might be left, he added, "I am leaving it to younger men."

He planned to do some traveling, perhaps a little salmon fishing. But
before that, he needed to reap the financial rewards of his laurels. In late
1926, after being home in Uranienborg in Norway for several months,
Amundsen returned to the United States to give lectures, although by
then he found them to be tedious fiscal obligations. He brought Hjalmar
Riiser-Larsen along, who alleviated much of the stress and in fact did most
of the talking.

Umberto Nobile had dutifully accepted Mussolini's orders and remained in the United States and was himself on a lecture circuit. What
followed was a bizarre period of nationalistic public mudslinging that in
many ways diminished and tarnished the historic achievement and the
reputations of the two central combatants.

In Nobile's lectures, and in interviews between them, he focused primarily on the Italian contributions, making it appear as if the entire enterprise had been conceived, organized, and executed by Italians. He took credit for coming up with the *Norge* flight idea himself, even though it was, in truth, Amundsen who had invited him to their "secret meeting" at Amundsen's home. Nobile went so far as to call both Amundsen and Ellsworth "merely passengers whom I took on board at Spitsbergen and left at Teller," and he said that they mostly slept during the entire flight. When an interviewer asked Nobile what he had to say about Amundsen referring to him as an "inferior officer," Nobile shot back: "I was the commander of the *Norge* and everybody on it, including Amundsen, was under my orders."

Amundsen referred to Nobile, on numerous occasions, as "nothing more than a hired pilot." He also suggested that Nobile's incompetence at navigation was what got them into trouble once they reached Alaska. He claimed that Nobile froze when they were in the mountains, and at one point "stood like a man in a trance," and later, "lost his head completely. With tears streaming down his face, he stood screaming . . ."

It was all deeply petty and personal, and unbecoming of the two men. But the airship had left the hangar, so to speak, and there would be no soft landing. Wrote one journalist at the time, summing up the juvenile tit for tat, "When the polar explorers landed in Teller there was honor and glory enough to go around, but if this quarrel continues there will be neither honor nor glory for anyone."

These words were nearly prophetic. The quarrel did continue, and even worsened. Amundsen and Ellsworth were both concerned by all the attention that Nobile was receiving in the United States. Nobile was invited to the White House by President Calvin Coolidge, and though it brought Amundsen and Ellsworth smug glee when they learned that Nobile's terrier Titina had soiled the president's rug, they were still jealous of the private audience. On this score they were overreacting, as both Amundsen and Ellsworth would eventually receive Congressional Gold Medals recognizing their contributions to the *Norge* transpolar flight. But in the moment, they believed that Nobile was cutting into their accolades as well as their profits.

When Amundsen and Ellsworth's book *First Crossing of the Polar Sea*[1] was published in the early days of 1927, it included no chapter from Nobile, although he had been contractually permitted to write a section on the aeronautical and technical aspects of the journey, which were of course his areas of expertise. But Amundsen and Ellsworth allowed the book to go to print without asking Nobile to submit anything, despite the subtitle being *"With Additional Chapters by Other Members of the Expedition."* But even worse, from Nobile's point of view, was that the book did not mention the official title of the expedition—the Amundsen-Ellsworth-Nobile Transpolar Flight—a title which he had fought for. They had excised his name from the expedition, as if he had not even been there.

Between traveling, lecturing, and dealing with personal matters, Roald Amundsen somehow managed to publish his memoir *My Life as an Explorer* in September of 1927. His decision to come out with the book, so soon in his retirement and while his emotions were still so raw, was questionable. He drafted the book hurriedly, mostly in his hotel room at the Waldorf-Astoria in New York in late 1926. Apparently, his editors at Doubleday were counting on his fame to sell copies rather than on the work's literary merits, which was unfortunate. The book was 277 pages of vindictive and slanderous score-settling, with an inordinate number of pages—nearly a third of the book—devoted to the *Norge* flight and mostly eviscerating Umberto Nobile. He excoriated Nobile as arrogant, a dandy and conceited boaster, a strutting peacock obsessed with uniforms who lacked discipline. Amundsen included a bizarre anecdote of a car ride he took in Rome, with Nobile at the wheel, which he called "The wildest ride I have ever taken in any craft." He intimated that Nobile's poor automobile driving suggested he was unsuited to operating an airship and

---

1 The book was first published in London in 1926 under the title *The First Flight Across the Polar Sea*; it was published the following year, 1927, under the title *First Crossing of the Polar Sea*, in New York by Doubleday.

ABOVE: Walter Wellman: political insider, explorer, impresario, 1910 (George Grantham Bain Collection, Library of Congress)

Wellman standing in the gondola (control car) of his airship *America,* in Oslo, Norway, 1906 (George Grantham Bain Collection, Library of Congress)

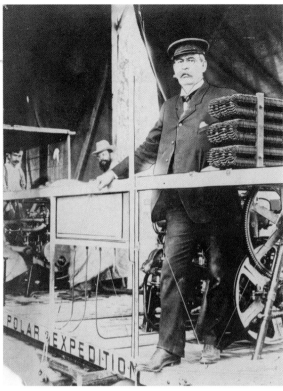

Wellman's airship station at Danes Island, September 1906. Foreground shows one of the propellers and the wicker basket that was to hold fuel tanks for the journey, but the basket failed during tests. (Anders Beer Wilse, Norsk Folkemuseum)

Camp Wellman and hangar, summer 1907. *The Illustrated London News* referred to it as "Mr. Wellman's scientific village in the Arctics." (Walter and Arthur Wellman Collection, National Air and Space Museum)

Wellman House and Virgo Harbor from shore, summer 1906. Wellman's proposed airship flight brought journalists and tourists to the remote Arctic island to witness history. (Anders Beer Wilse, Norsk Folkemuseum)

Wellman's airship *America* leaving the hangar at
Virgo Bay, September 2, 1907 (George Grantham
Bain Collection, Library of Congress)

Wellman and the crew of the 1907 Wellman
Chicago Record-Herald Polar Expedition. (*From left
to right, front*) Melvin Vaniman, Felix Riesenberg,
Henry E. Hersey. (*Back, standing*) Wellman.
(Anders Beer Wilse, Norsk Folkemuseum)

America at polar pack,
August 1909
(National Air and Space
Museum Archives)

America rescued by the Norwegian vessel *Farm*, August 1909 (National Air and Space Museum Archives)

America about to be rescued by the *Trent*, Atlantic Ocean, October 1910 (George Grantham Bain Collection, Library of Congress)

A dapper Roald Amundsen in bowler hat, 1926 (National Library of Norway)

Amundsen posing in polar attire, Nome, Alaska, 1923 (Norwegian Polar Institute)

Amundsen standing next to a Dornier-Wals "flying boat," Ny Ålesund, Svalbard, May 1925 (National Library of Norway)

Lincoln Ellsworth inspecting the Dornier-Wals prior to departure, Ny Ålesund, May 1925. Ellsworth's millionaire father provided the funding to build the two state-of-the-art aircraft, *N24* and *N25*. (National Library of Norway)

N24 on the ice at Ny Ålesund, ready for departure, May 21, 1925 (National Library of Norway)

The *Norge* in flight over Ekeberg, Norway, en route to Svalbard, April 14, 1926. Men are visible in the engine cars on either side of the control car. (National Library of Norway)

The *Norge* is walked into the hangar at Ny Ålesund, May 7, 1926. Preparations were nearly complete, and Umberto Nobile was anxious to start, hoping to beat Commander Richard E. Byrd in a race to the North Pole. (National Library of Norway)

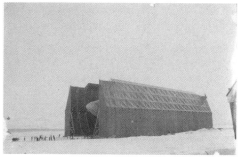

ABOVE: The *Norge* safely ensconced in the hangar at Ny Ålesund. Amundsen awaited good weather for the flight across the polar sea. (Norwegian Polar Institute)

While waiting for departure, Ellsworth snowshoed around Kings Bay (Ny Ålesund) to stay fit, while Amundsen (*not pictured here*) Nordic skied daily. (National Library of Norway)

Just before departure from Kings Bay, *May 11, 1926*
(National Library of Norway)

LEFT: The *Norge* hovering near the mooring mast at Kings Bay, May 1926. The mooring mast is still there today.
(National Library of Norway)

BELOW: The *Norge* departs, carrying Amundsen, Ellsworth, Nobile, and thirteen crewmembers, May 11, 1926.
(National Library of Norway)

First aerial view of the ice around the North Pole, taken from the *Norge* on May 12, 1926 (National Library of Norway)

RIGHT: The *Norge* heading into a storm. The flight from the North Pole to Alaska was arduous and death-defying. (National Library of Norway)

BELOW: The *Norge* making a semi-controlled emergency landing at Teller, Alaska, after a historic flight of seventy-one hours, covering twenty-seven-hundred miles, May 14, 1926. The explorers had disproved the existence of a landmass near the North Pole. (Colaimages/Alamy)

The beleaguered airship *Norge* after being deflated at Teller. Nobile was able to salvage the engines and some of the other airship parts. (Alamy)

Ellsworth (*left*) and Amundsen posing in Nome after their triumphant transpolar flight. They appear happy, but a rift with Nobile has already begun and is about to escalate into a public battle of words. (National Library of Norway)

Nobile in formal military dress, with plans and maps before the *Italia* departure to the Arctic. (Süeddeutsche Zeitung Photo/ Alamy)

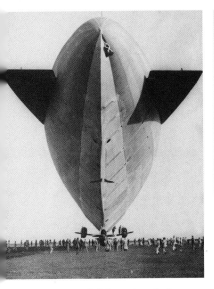

ABOVE LEFT: Nobile's airship *Italia* from the stern (showing prominent lateral tail fins) at airfield in Milan, preparing to depart for Spitsbergen, Norway, April 1928. Because the city of Milan was a major sponsor, many citizens came to see the airship while it was there. (Süeddeutsche Zeitung Photo/Alamy)

ABOVE RIGHT: The beautiful, streamlined *Italia* in flight (Süeddeutsche Zeitung Photo/Alamy)

MIDDLE RIGHT: The *Italia* flying over the north coast of Spitsbergen. It arrived at Kings Bay on May 6, 1928, after a series of flights covering more than twenty-four-hundred miles from Italy. (Süeddeutsche Zeitung Photo/Alamy)

When Amundsen heard that Nobile and the *Italia* were missing, he proclaimed, "I am ready to start on a search for the *Italia* at once!" Good to his word, despite their bad blood, Amundsen boarded a Latham 47 sea plane (*here about to leave Tromsø, Norway*) and left with a crew of five on June 18, 1928. (Norwegian Polar Institute)

On June 23, 1928, Swedish pilot Einar Lundborg rescued Nobile and his dog, Titina, from the Arctic ice above Svalbard. Begrudgingly, Nobile agreed, at Lundborg's insistence, to be the first man rescued. It was a decision that would haunt Nobile for the rest of his life. (Einar Lundborg and Swedish Airforce Museum)

After delivering Nobile to safety at Rossøya Island (*northern Svalbard*), Lundborg returned to the survivors of the Red Tent but crashed his Fokker during landing, flipping the ski plane upside down. Now, he would need to be rescued along with the other *Italia* survivors. (Einar Lundborg and the Swedish Airforce Museum)

The survivors of the *Italia* crash dyed their tent red for visibility from the air. (*Seated to the left*) Professor Francis Behounek prepared polar bear meat for dinner. Alfredo Viglieri (*at right*) is loading lifeboats with gear as the ice is disintegrating around their camp. To the right of Viglieri is Giuseppe Biagi's radio mast. (Einar Lundborg and Swedish Airforce Museum)

Behounek sits on ice at the Red Tent camp, cooking polar bear meat. (*Back left*) The carcass of the polar bear shot by Finn Malmgren is visible. (Einar Lundborg and Swedish Airforce Museum)

The Red Tent was moved as ice deteriorated. In front of the Red Tent is the tin fuel can used to melt snow for drinking water. In foreground are collapsible boats loaded with gear for emergency water departure. (Einar Lundborg and Swedish Airforce Museum)

The *Italia* disaster survivors sitting on underside of the wing of Lundborg's upside down Fokker, with clothes drying above them. (*Left to right*) Behounek, Biagi, Viglieri, and Natale Cecioni (*with leg splinted using ski poles*). (Superstock/Alamy)

The icebreaker *Krassin* finally arrives to within one hundred yards of the *Italia* survivors at the Red Tent on July 12, 1928. Biagi's radio mast, which was instrumental in gaining contact with the outside world, is to the far right of the frame. (iStock)

Krassin crew and survivors dismantling the Red Tent, July 12, 1928. One of the last things radio operator Biagi did was retrieve from the center tentpole the image of *Madonna of Loreto*, which had sustained them through their long ordeal. (Süeddeutsche Zeitung Photo/Alamy)

The *Italia* survivors in front of the train that returned them home to Italy. (*Left to right*) Ettore Pedret (*holding Titina*), Felice Trojani, Guiseppi Biagi, Viglieri, Nobile (*with cane*), Cecioni (*on crutches*), and Behounek. (Süeddeutsche Zeitung Photo/Alamy)

Pathfinder 1 undergoing flight testing operations outside at Moffett Field (LTA Research)

was emblematic of his character: "His whole performance on the trip," he wrote, "was evidence of his extreme nervousness, erratic nature, and lack of balanced judgement."

In this case, the lack of balanced judgement was Amundsen's. The book was poorly reviewed, did not sell well, and caused a number of close friends and associates to wonder whether all was right with his mind. He had suffered depression before, and the memoir appeared to have been written while he was in a dark and brooding place. Country-man Fridtjof Nansen, concerned by the book and for Amundsen's mental health, wrote to a mutual colleague, "I think he is suffering from some sort of mental confusion, a sort of clinical nervous disability which has in fact been evident in many ways."

Amundsen, by every measure, had accomplished a towering career as an explorer. But that life, the constant travel to the remotest places on the planet, the hardships endured, the more than two decades of planning and exploring and lecturing and writing, had left him distant, paranoid, and unsuited to the mundane pace of normal life. He was not a man content with an idle life. His closest and most enduring relationships had been with fellow explorers, and the few romantic connections he nurtured tended to be with married women, probably because they required only the scant commitment that the global explorer could give.

In late 1927, Amundsen was in New York between lectures in support of his memoir, when he abruptly canceled the remaining dates on his tour and departed by steamship for his homeland. This was further erratic behavior, and there were rumors swirling that the reason for his hurried escape to Norway—in addition to salving his self-inflicted literary wounds—was to rendezvous at Uranienborg with Bess Magids, a striking married woman of thirty he had met five years before aboard the steam-ship *Victoria,* traveling from Seattle to Nome.

Umberto Nobile, meanwhile, had concluded his successful propagan-distic lecture tour of the United States—which included events in New York, Washington, D.C., Chicago, San Francisco, and Los Angeles—and

returned to a hero's welcome in Italy. When he had left Rome's Ciampino airfield at the helm of the *Norge* in April of 1926, Nobile had been
a relatively obscure Italian airship designer and engineer; he came back
to the same airfield four months later, "in an open carriage with a cavalry escort," through throngs of cheering countrymen, wearing a dapper,
bleached-white dress uniform. He was now a celebrated polar explorer
and a recipient of the U.S. Congressional Gold Medal. He had done Mussolini's bidding promoting Fascist Italy's exploits in aeronautics and had
raised their nation's global stature, enough so that the Prime Minister
awarded Nobile the prestigious Military Order of Savoy and promoted
him to general.

Umberto Nobile reveled in the accolades and awards, but he wanted
more. "There were still 1,500,000 square miles of unexplored territory
within the Arctic circle," he wrote of his new idea, with the most interesting areas to explore "along the coasts of Greenland, Siberia, and Canada . . . to undertake an ambitious programme of scientific research in the
fields of oceanography, terrestrial magnetism, gravity, atmospheric electricity and radioactivity, and Polar biology." He would show the world
that he did not need Roald Amundsen, that the Italians could go it alone.

He would, or course, need Mussolini's permission to enact his ambitious new plan. And he was going to need a new airship.

PART THREE

THE *ITALIA*

Disaster at the
Top of the World

25

BIRTH OF THE *ITALIA*

UMBERTO NOBILE SAT NERVOUSLY ACROSS FROM BENITO MUS-solini in the Prime Minister's ornate office. He needed this meeting to go well. Nobile had asked for the meeting to put forward a proposal for another dirigible polar flight, this one even more ambitious in scope and scale, and one that would be an entirely Italian expedition, flying the Italian rather than the Norwegian flag. Mussolini, well aware of the propaganda windfall of the *Norge* transpolar flight, had been intrigued enough to consent to see Nobile.

Nobile explained to Mussolini that the *Norge* flight across the Arctic Ocean, as successful as it was, had been essentially a flyover, primarily aeronautical in character. It had led to important geographical understanding of the area between the North Pole and Alaska, but beyond that, little science had been achieved. With a modified airship, one with more range, lift, and better ballast for hovering slowly and making observations, there remained much science to perform in Arctic geography, meteorology, and oceanography. In the right conditions, Nobile said, men could even be lowered down to the ice near the North Pole, and at other locations not yet sufficiently studied, for protracted on-ice research. And there were already mooring masts and existing hangars they could use, at Vadsø and Kings Bay. It would be a shame to let them languish, or worse, be used by other nations first.

Mussolini straightened in his chair and shuffled a stack of papers. He tapped his fingers on a leather-bound notebook in front of him, his lips rising at the edges in a slight smile. "Perhaps it would be better not to tempt fate a second time," he said. Mussolini understood the inherent dangers of all dirigible flight, and lighter-than-air flight in the Arctic regions further exacerbated those risks. They were also very expensive. Nobile started to speak, hoping to argue his case with a list of reasons that the idea made sense for Italy, when Mussolini stood, indicating that the meeting was over. "Still," he said as Nobile rose to depart, "I recognize the scientific importance of the idea. We will talk about it again next week."

They did not talk about it the next week, but Nobile soon received word from Italo Balbo, undersecretary at the Air Ministry, that if Nobile could secure some private funding to help defray costs to the Italian government, Mussolini would sign off on the proposal, which included allowing Nobile to use the *N4,* the last of the *N*-series. It was a sister ship to *Norge* but would require numerous modifications for Arctic conditions. Mussolini offered an expedition vessel from the Italian Navy for transport and support, as well as flight officers and crew. Nobile parlayed his recent notoriety, and his new rank as general, to get meetings with the Italian Geographical Society at their Palazzetto Mattei headquarters. They were intrigued by the scientific possibilities of dirigible use in the Arctic and enthusiastically agreed to sponsor the expedition, lending their name and global gravitas. With their aegis obtained, Nobile appealed to the City of Milan for the funding required to pay for the construction and insurance of the airship, as well as to pay the flight crew. By the time all of the agreements had been signed by the supporting parties, it was December 1927.

Nobile was pleased with the arrangements but there was much work to be done. He had wanted to build a much larger Arctic airship from scratch, but as the *N4* was all Mussolini had offered, it would have to do. Nobile set a tentative departure date from Italy to Kings Bay, which would be the staging area for the proposed flights, of April 1928. That was only a few months away—and he still needed to retrofit the airship and hire a crew.

Nobile had learned much about the perils of Arctic air travel with the *Norge,* and at the Italian State Airship Factory in Rome, he directed a number of modifications to the *N4.* To thwart the problem of propeller-flung ice tearing the envelope, he reinforced the varnished canvas with three-ply rubberized fabric. He also strengthened the covering of the keel with multiple extra layers of material, laid down in diagonal strips, thickening the original cloth. Internally, he fortified the rigid end of the prow for better performance against the wind, especially while attached to a mooring mast. He added protective hoods on the gas valves to prevent ice formations from jamming them open, which risked loss of hydrogen, or freezing them shut, which would prevent them from being opened to valve off gas. And, recalling how cold the cabin of the *Norge* had become, sometimes to the point that the portals and windows had iced over from condensation from the inside, Nobile insulated the commander's cabin by lining it with canvas inside and out.

Additionally, Nobile installed an upgraded "wireless plant" for communications, with a Marconi-Bellini-Tosi direction finder (radiogoniometer) to determine the direction to, or bearing of, a land-based radio transmitter. Nobile also made sure the wireless plant had stocks of batteries, backup shortwave radios, and an emergency transmitting apparatus. To combat the antenna icing that had plagued him on the *Norge* flight, Nobile devised a system whereby the antenna could quickly be retracted into the cabin during fog or freezing rain conditions.

Nobile compartmentalized the keel to provide a storage area for exploration equipment, including snowshoes and skis, inflatable rubber boats, sledges for man-hauling, tents and cookstoves, firearms, ammunition, and fuel cans. A section was devoted to spare airship accessories: extra propellers, engine parts, and tools. The keel also had an area for sleeping, with hammocks suspended crosswise above the catwalk, high enough so as not to impede men moving along it to do their work.

When the airship was finished, it was 347 feet long, 64 feet in diameter, with a capacity for 654,000 cubic feet of hydrogen. The modernized aircraft, propelled by three Maybach diesel engines, possessed a top speed of seventy miles per hour, and bore a fascist symbol painted on the side of the control cabin. Nobile was impressed by what he and his engineers had accomplished, boasting that the airship "had reached a point of perfection that placed it in the front rank of dirigibles of similar cubic capacity that had recently been constructed, whether in Italy or Germany." Nobile named it, fittingly, the *Italia*.

With *Italia* ready, Nobile began narrowing down the crew from a list of hundreds of applicants. His initial choices gave priority to members of the *Norge* flight, so he asked the experienced Renato Alessandrini (rigger), Vincenzo Pomella (motor mechanic), Attilio Caratti (motor mechanic), and Ettore Arduino (chief motor engineer) if they would come again to the Arctic. Nobile confided in them that he believed the dangers of the new venture to be greater than those of the 1926 *Norge* expedition, but was proud when "All four accepted enthusiastically, without discussion, without making stipulations or conditions of any kind." So it was that all the Italian contingent on the *Norge* would also accompany Nobile on the *Italia*. To this able and experienced group of engine operators and mechanics he added a man named Calisto Ciocca (motor mechanic), with whom he had worked for years constructing and test flying airships. Nobile wanted Ciocca as an extra man to ensure adequate rest during the long flights, so that men could provide relief to one another.

Nobile's only hesitation with the group from the *Norge* was Alessandrini, who, after the *Norge* flight, had developed a considerable girth. Nobile joked with him, saying, "You'll not be able to get through the porthole!" It had been Alessandrini, in fact, who numerous times had crawled along the steel tubes of the catwalks and then squirreled through the tight bow porthole and up the steel ladder to inspect the external gas valves. Alessandrini had just grinned, assuring Nobile that he'd taper down and would still be able to get through.

Content that the *Italia*'s engines would be in the best possible hands, Nobile next sought two superb officers to navigate and also to take astronomical observations. He deferred to the Italian airship admiralty, and they immediately recommended Adalberto Mariano (Commander, First Officer) and Filippo Zappi (Commander), each of whom he interviewed and accepted without reservations. The admiralty offered a third officer, and Nobile picked a man named Alfredo Viglieri (Lieutenant Commander), intending to employ him with assisting in navigation and also to periodically take the helm. Nobile added another man, Felice Trojani (Engineer), whom he had known for twelve years and who had worked with him designing and building airships. Wireless operators would be crucial, and Nobile chose Ettore Pedretti and Giuseppe Biagi, two petty officers "well worthy to join the elect group of veterans of the *Norge*."

As it was intended to be a journey of science and exploration, Nobile wanted scientific scholars of the highest order aboard. His first pick was an Italian named Aldo Pontremoli, a professor of physics from the University of Milan, who had been recommended to him by eminent scientists in Europe and the United States. Pontremoli had been appointed, in 1926, to a newly created group of theoretical physicists along with Enrico Fermi (later renowned for being the creator of the world's first nuclear reactor and a member of the Manhattan Project, as well as the namesake of the element fermium). Pontremoli's cause was helped because, as Nobile put it, he was "not new to aeronautical ventures." Pontremoli would be responsible for taking measurements of the Earth's magnetic field and cosmic rays.

Nobile's next two choices were the only cases in which he diverged from having an entirely Italian crew, and he had good reasons for both. Swedish scientist Finn Malmgren had distinguished himself as a meteorologist on the *Norge,* and Nobile had found the University of Uppsala professor a vigorous young man in his early thirties, "rather short, fair, with blue eyes and a keen glance." He had a superb mind, but he was also rugged and tough, having been with Roald Amundsen from 1922 to 1925 in the Arctic on the *Maud* Expedition. Nobile liked the idea of taking someone with skiing, snowshoeing, and ice travel experience, and he, of course, knew him personally already. Though he was Scandinavian, Nobile harbored no ill will against Malmgren. The Swede was also straightforward, with a good sense of humor, which was useful in the tight confines of an airship. When Nobile approached him with the offer, they talked about the proposed itinerary, and the inherent weather difficulties they were likely to encounter. "For my part," Malmgren told Nobile, "I think the probability of success is about 50 percent." Then he smiled and added, "But our chances of saving ourselves are, of course, greater."

The last scientist Nobile invited was Dr. Francis Behounek, of the Prague Wireless Institute. Nobile had met the Czechoslovakian Behounek at Kings Bay in 1926, while waiting to depart on the *Norge*. At the time, Behounek had been performing experiments in the electricity and radioactivity above the Arctic Circle, and Nobile had been impressed with the man's energy, enthusiasm, and intellect. Professor Behounek also had been highly recommended by none other than Marie Curie, who by that time had won two Nobel Prizes—one in physics and one in chemistry—so her endorsement carried considerable weight.

After months of intense work, Nobile had his crew and airship ready. In late March of 1928, he sent the naval vessel *Città di Milano,* loaded with gear, provisions, and the men Mussolini had promised as ground/support crew, to Kings Bay. Nobile had arranged that at the same time another ship, the 150-ton sealer *Hobby,* would depart Tromsø, Norway, for Kings Bay as well, bringing along a master builder, a new canvas covering for the hangar, and hundreds of tons of materials as backup should the *Città di Milano* encounter bad ice or weather conditions and fail to arrive. Also aboard the *Hobby* was a small party of Italian ski troops or "Alpini," led by Gennaro Sora; they were included as a rescue party. Nobile's brother Amedeo would go along on the *Hobby* and manage things on the ground at Kings Bay.

Now, it was time for Nobile to convene his crew and prepare to fly to Kings Bay and into the Arctic beyond. But first, he had an important audience with Pope Pius XI.

26

BLESSED

UMBERTO NOBILE AND HIS CREW STOOD, QUIET AND RESPECTFUL, in the great hall of the Vatican. Pope Pius XI entered the hall, robed and wearing wire-rimmed spectacles, and, as he strolled slowly past, was introduced to each member personally. Each man stepped forward, said his name, then fell back in line, lowering their heads in reverence. Then their pontiff began speaking, his voice solemn but serene, in a tone, as Nobile heard it, that "a father might speak to his children who were setting out on a noble but dangerous venture." His paternal words echoed lightly through the hall as he bestowed on them the courage and conviction they would need to triumph.

Nobile had met with the Holy Father numerous times in the past few years, and they shared a deep connection. Pope Pius XI had taken a keen interest in the *Norge* flight and the upcoming *Italia* expedition, and during their meetings, Nobile had learned that the pontiff was an ardent student of polar exploration and history, and was himself an avid mountaineer, having spent a great deal of time in the Alps, reaching the summits of the Matterhorn and Mont Blanc, both difficult and dangerous ascents. Nobile had been humbled by his personal attention and advice.

Now, the Holy Father, raising his open arms, addressed the entire crew and prayed for them, saying, "In all human undertakings, at a given moment, a higher force intervenes, to turn the scale of human endeavor for good or for ill." Then he produced a large, hand-hewn oak cross, which he had made himself, and presented it to Nobile, imploring him to carry it on the *Italia* and plant it at the North Pole. Nobile took the cross and studied it. Its top had been hollowed out, and inside the cavity was a rolled parchment. Nobile took out the roll, which read as follows:

> The emblem of our Lord Jesus Christ had today, the 31st of March, 1928, been entrusted with dedicatory prayers by His Vicar, Pius XI, Pontifex Maximus, to Umberto Nobile and his companions, on the eve of their aerial journey at the charges of the City of Milan, to be dropped by the leader of the expedition, flying for the second time over the Pole; thus to consecrate the summit of the world.
>
> Pope Pius XI then blessed them, each in turn. He paused and spoke to Nobile, eyeing the cross and smiling. "And like all crosses," he said, "this one will be heavy to carry."

During the next few weeks, Nobile finalized details in Rome, spending as much time with his wife, Carlotta, and ten-year-old daughter, Maria, as he could, though as usual his home time was limited. Nobile was an incessant worker, often putting in eighteen-hour days, and though it was difficult on his small family, they had become resigned to it. But when he was about to leave on long airship journeys, Carlotta suffered insomnia. Nobile was always honest with her about the dangers involved in dirigible flight, explaining that while every detail had been planned and tested, there were always risks. She had also read many accounts in the newspapers chronicling airship disasters. The truth was—and Carlotta knew this in her heart—every time her husband Umberto Nobile climbed into a dirigible, there was a chance she might never see him again. But she was strong, self-controlled, and even-tempered, and she listened attentively to the counsel Nobile gave her about raising Maria in the event that he did not return.

Then, just as she had when he departed on the *Norge,* she kissed him gently and said, "Go away happy!" He found her words deeply comforting.

Before departing for Milan, which, as the sponsoring city, was the official starting point of the expedition, Nobile had a formal farewell audience
with King Victor Emmanuel III. The King was intrigued and inspired by
the proposed journey, especially since he had twice been to the Arctic as
a young boy. He was inquisitive and supportive, offering "fervent good
wishes for the success of this new enterprise."

On April 11, a few days before leaving for Milan where *Italia* was waiting, Nobile had a final meeting with Mussolini. Nobile unfolded a large
map on the prime minister's desk and traced the entire route, from Milan
to the German air station at Stolp, near the Baltic Sea, then to Vadsø,
Norway, and across the Barents Sea to Kings Bay, Spitsbergen. After that
nearly three-thousand-mile journey, from Kings Bay they would make
flights to Severnaya Zemlya north of Siberia, the Franz Josef Land archipelago, over the north coast of Greenland, and finally to the North Pole
and back. It was by far the most ambitious Arctic aerial itinerary ever
proposed. Mussolini nodded enthusiastically, then he said, "The enterprise
is not one of those destined to strike the popular imagination. But it will
attract the attention of the scientific world. I see already that there is a
great deal of interest in it abroad." With that, he wished Nobile good luck
and bade him farewell.

In Milan, on the night before departure, Umberto Nobile addressed a
gathering of Milanese dignitaries, citizens, and reporters. The *Italia* had
caused a lot of excitement while stationed in the city at its specially built
hangar, and now people wanted to hear Nobile speak about the expedition. He had also been reading news stories in the papers, some suggesting
that the journey was foolhardy and would end badly. Nobile wished to
address those naysayers. "We are quite aware that our venture is difficult
and dangerous," he told those assembled, "even more so than that of 1926."
He paused, looking directly into the eyes of those in the front row of the
crowd, and then dramatically cast his eyes skyward. "But it is this very difficulty which attracts us. That is precisely why we are doing it. Had it been
safe and easy, other people would already have preceded us." Everyone

applauded, and Nobile left to go pack the last of his personal belongings for the first leg of the flight.

To the degree that a 350-foot dirigible can slink away into the night, the *Italia* did just that. Nobile told only his men and ground crew of their secret departure at 2:00 A.M. on the morning of April 15, not wanting "a troublesome crowd on the maneuvering ground and in the hangar" as they were leaving. It worked, and as most of Milan slept, they flew away into dark skies despite an unfavorable weather forecast all the way to Stolp. As they ascended into bumpy air, the Swedish meteorologist Finn Malmgren shouted to Nobile, "We shall have some excitement!" They had previously discussed waiting longer, but the weather looked stormy for at least a week, and Nobile knew that the best Arctic flying period—early spring—was upon them, providing nearly twenty-four hours of continuous daylight and less fog. So, he had made the decision to start on the first leg of the shakedown flight, a 750-mile journey from the Valley of the Po River in Italy, skirting the eastern massif of the Alps, then north over central Europe to Stolp for refueling before continuing on to Vadsø, Norway.

It did not take long to encounter the excitement Finn Malmgren had predicted. Just past Trieste, Italy, they were slammed by a 40-mile-per-hour gust that warped and distended the *Italia*'s internal frame and damaged the left horizontal fin. Nobile's primary concern was that one of the steel tubes inside the fin had been exposed, and he worried it might protrude further and rip the envelope. He personally examined the fin, determined the damage to be localized, and decided to press on and repair it later. They flew on through darkening gray skies, receiving a weather report from Prague: "Dismal, with a nasty drizzle, and flashes of lightning, with 25–35 mph winds, and gusts much higher." The news was unsettling, but Nobile was not about to turn around now.

As they approached the Sudetes Mountains on the Czechoslovakian border, Nobile saw that their route was blocked by masses of dark clouds. After consulting with Malmgren, they chose to proceed with caution, as they were nearing a mountain pass. With mountains rising all around them, a sudden hailstorm struck them, coupled with bursts of lightning, the frightening flashes punctuated by pelting hail that sounded like drum-beats on the envelope. "It was a terrible moment," Nobile recorded of the atmospheric event. "The storm was . . . on all sides . . . lightning to right, lightning to left, in front of us, behind us, accompanied by deafening peals of thunder. It seemed as if there were no way out." Everyone on board understood the danger. One lightning strike could set afire or explode the *Italia*. Yet, to Nobile's credit in choosing his crew, all on board remained calm and composed, concentrating on their duties.

Nobile ordered descent to get below the lightning, and for the next hour they sailed through mountains and storm, swerving and weaving their way in a white-knuckle ride until they had left the mountains and passed into clearer skies. The next weather report was much improved, and they cruised through calmer air and good visibility, easing down to the airfield at Stolp on the early morning of April 16. Nobile and his engineers inspected the *Italia* and were amazed by the damage the storm had wrought: the upper tail fin had also been shattered, and the propellers were pocked and shredded where hail had struck them.

Nobile ordered part of the crew to rest while others began working on repairs. He calculated they would be grounded for a week. He wired to Rome requesting spare and replacement airship parts and made inquiries about the progress of the steamships *Hobby* and *Città di Milano*. The *Hobby* had already made it to Kings Bay with men and materials, and the hangar and mooring mast were ready. Nobile ordered the *Città di Milano* to proceed there from Tromsø immediately. Nobile also sent a telegram to Kings Bay asking the mining company there for men to assist in landing *Italia* when it arrived, if needed. The Norwegians there had helped with *Norge* two years before and consented without question.

It was nearly two weeks before the repairs and weather made it possible for Nobile, his eighteen crew members, and little Titina to be airborne again. On April 30, weather reports were decent along the northern Scandinavian coastline, and moderate from there to Spitsbergen, with high winds and snow predicted over the Barents Sea. The Tromsø Geophysical Institute advised Nobile to start immediately, adding: "Such a situation is as good as one can hope for at this time of year. There are uncertainties and risks, but these are inevitable." After what he and the *Italia* had recently been through, this last bit of information went without saying.

On May 3, Nobile and Malmgren finally agreed the local winds had subsided enough to ensure safe departure, and they nosed into predawn skies and pushed into a strong headwind from the north. Nobile ran *Italia* at full speed, and they made the 410 miles to Stockholm in eight hours. As they neared the city, crewmen peering from the airship could see Swedish naval seaplanes flying along in escort, and then, as they flew over one neighborhood, Malmgren yelled out, "There is the house!" He could see the house where his mother lived, and Nobile agreed to circle around and pass low enough for Malmgren to wave to her and toss out a letter for his family. Malmgren swelled with pride at his position among the few dirigible meteorologists in the world, and especially to be aboard the *Italia,* whose ambitious itinerary had garnered so much international interest. He pictured his mother telling her neighbors, "That's my son Finn up there," and pointing to the airship as it flew magisterially away toward Finland and beyond, to the roof of the world.

They encountered snow and ice over northern Scandinavia and also received a weather report that gave Nobile and Malmgren pause, predicting dangerously strong winds headed their way. "We consider it advisable," the report read, "to increase speed of the ship in order to utilize present favorable landing conditions." With that, Nobile had all engines brought up to full speed and, at times reaching 70 miles per hour, they made Vadsø, docking at the mooring mast just ahead of the coming gale.

Nobile's original plan had been to remain in Vadsø only long enough to refuel and take on more vital hydrogen gas, but once these tasks were completed the winds came on with such ferocity that the *Italia* spun around the rotating masthead 360 degrees, whirling like a pinwheel. It was so dizzying and dangerous that Nobile kept the motors and propellers running for hours, pointing the ship's nose into the wind to reduce the spin, which was threatening to tear the ship loose from its mooring mast. The men on board cringed as gusts blasted them from the north at 40 to 50 miles per hour, and they could hear the internal framework groaning and wailing until one of the steel tubes at the stern buckled, snapping with the sound of a gunshot. Nobile called for a team to repair it, which they did. Throughout the day they remained under the siege of the storm, with rain running down the envelope in sheets and sloshing into the canvas-covered cabin, soaking everyone and everything.

Finally, as evening approached, the winds began to subside. Weather reports from Tromsø, and from Bear Island (Bjørnøya) in the middle of the Barents Sea, were not ideal but Nobile did not want to remain on the mainland any longer, lest they end up stuck there and the entire enterprise get canceled for another year. Nobile called all men to their stations and at 8:30 P.M., May 5, they lifted into the air and bore north toward Spitsbergen. They battled a strong headwind through the night, and by sunrise could see the high vertical cliffs of Bear Island, a remote, lone island of just seventy square miles where a small meteorological station had been installed by Norway in 1919. The island was blanketed with fresh snow. Winds were light and visibility was good enough that Malmgren convinced Nobile to fly low over the weather station and wave to the two isolated physicists working as meteorologists at one of the farthest-flung outposts in the world. Nobile had greatly appreciated their recent weather reports and obliged, banking a slow turn over the small observatory and dipping the nose in a nod, then pushed northward.

Between Bear Island and Spitsbergen, the weather deteriorated further, and as Nobile wrote in his log, "We found ourselves in the head of a cyclone advancing from Iceland, which set up violent southeast winds [40–60 mph] on our course." Buffeted by intense side winds, they were blown off course and ran directly into thick cloud masses and a driving snowstorm as they reached the serrated southern coast of the Spitsbergen archipelago, still nearly two hundred miles from safety at Kings Bay. In the midst of the tumult, Nobile considered his options. Checking the gauges and consulting with the mechanics, Nobile determined they had fuel enough for three days, so he was willing to fly around until the storm abated. But Arduino, the chief motor engineer, reported that one of the engines had cut out and could not be repaired midair. They would need to land soon.

Radio operator Biagi contacted Kings Bay, which reported strong gusts on the ground, but there was some good news: the ship *Città di Milano* had arrived safely carrying 166 crew, reserve engines, and many backup parts for the airship. The weather began to clear, and Nobile was able to see that they were flying up the strait between Prince Charles Foreland, Spitsbergen's westernmost island, and the Svalbard mainland. Flanked on either side by high mountains, Nobile told Biagi to alert Kings Bay they were coming, and to have men ready at the mooring mast and hangar. At noon on May 6, Mariano, manning the controls, called out that he could see the houses and mining buildings, and the *Città di Milano,* decked out in Italian flags, at anchor in the Kings Fjord. The smaller *Hobby* was there too.

As they approached from the south and circled over the bay, Nobile had Biagi radio Captain Romagna of the *Città di Milano* requesting fifty sailors to assist the men from the *Hobby* on the ground to bring them in. The response he received was perplexing and maddening: "My men cannot be spared from their duties," Captain Romagna called back. Nobile barked back that no duties could be more important than landing the *Italia* at this moment, but Captain Romagna said he would have to radio Rome to get official orders from the naval admiralty there. Disgusted, Nobile cut communication with Captain Romagna and got out his landing megaphone. "Get me the miners!" he belted down to the men already assembled at the mooring mast 150 feet below. In a few minutes, Norwegians employed by the mining company came trotting over to join the *Hobby* crew. All held their hands up, waiting for lines to be tossed down.

Nobile directed maneuvers by megaphone. At first, he lowered the mooring cable and had *Italia* secured to the mooring mast, but during the process, a couple of the airship's metal nose cone tubes broke. This weakened the point of attachment, and Nobile worried that a gust might tear the ship loose from the masthead, so he detached from the mooring mast and, during a lull, directed the groundcrew of one hundred men as they guided the beleaguered but airworthy *Italia* safely into the hangar.

When Nobile finally stepped out of the control car onto the ground at Kings Bay, he had been awake and actively engaged in various duties— other than during a few short catnaps—for eighty-two hours. Standing on wobbly legs, Nobile rubbed his fatigued eyes and took a long look at the plucky airship as his men dutifully valved off hydrogen gas and secured her with sandbags and lines. He and the crew had flown the *Italia* a total of 2,187 miles from Stolp, with the last leg of eight hundred miles taking just over fifteen hours, at an average speed of 50 miles per hour. As he turned from *Italia* and walked toward the headquarters to get some sleep, he fondly recalled the audience with Pope Pius XI, shaking his head in pride at what he and his Italian dirigible and crew had already endured just to reach their actual expedition starting point. Truly, they were blessed.

27

ARCTIC FORAYS

UMBERTO NOBILE DID NOT REST LONG BEFORE HE RETURNED TO the hangar to assess the condition of *Italia*. Two of the engines needed on-site overhaul, and he set mechanics straight to work on those, and ordered other men to begin refueling the petrol tanks, refilling the hydrogen gas cells, and checking and rechecking all of the gauges and instruments. The engine that had quit during their recent flight over the Barents Sea needed to be replaced. Fortunately, backup engines had been brought along on the *Città di Milano*, and he instructed men to convey them to the hangar, along with hundreds of pounds of emergency gear and provisions needed for their upcoming Arctic exploratory flights.

The work was exacting and time-consuming, partly because the *Città di Milano* had to anchor slightly offshore, along a band of encroaching ice instead of at the wharf, which meant that the heavy engine, parts, and gear had to be transported about a mile and a half through deep snow. After four days of constant work, *Italia* was once again deemed ready to fly. Nobile was satisfied with the condition of his airship, and also with the emergency gear he had ordered specially made in Norway. Before departing Italy, Nobile had traveled to Oslo and consulted with Fridtjof Nansen about the best kinds of fur clothing, tents, skis, snowshoes, footwear, and rations for travel and survival on the polar surface. Nobile believed that it might be possible, in perfectly calm conditions, to lower men to the ice for a few days of study and retrieve them later. He had invented an ingenious pneumatic hoist and basket for the purpose. The men would need the best Arctic equipment for this dangerous duty, especially if the weather changed and stranded them on the ice for days, weeks, or even longer.

But Nobile had also been preparing for emergency. It was not hard to imagine that some catastrophe might befall them—failed engines, icing, ruptured envelope, the list was nearly endless—and they would find themselves on the ice. Should such a calamity occur, he wanted to be prepared for what might be an extended time on the ice, and potentially lengthy travel to return to land. Nansen had recommended reindeer-hide sleeping bags, reindeer-skin boots filled with dry grass for insulation (three pairs per man), and silk four-person tents. Nobile even obtained his pemmican—the polar explorer staple typically made of pulverized and dried meat, fat, fruit, and vegetables—from the company that Nansen had recommended.

Nobile inspected the crates of Arctic gear at the hangar and supervised the loading of it onto the *Italia*. For garments, he ordered each man to wear or bring woolen long underwear, knee-high woolen socks, a woolen mountaineer's hat, thick gauntleted leather gloves, and a three-piece lambswool flying suit of parka, trousers, and hood.

During all of these preparations, Nobile had discussed his objectives with his leadership team of Commanders Mariano, Zappi, Lieutenant Commander Viglieri, and meteorologist Malmgren. He reiterated his plans to make four-to-five major exploratory flights over the next month, including Franz Josef Land, Siberia, Greenland's north coast, Severnaya Zemlya, and the North Pole. Nobile underscored that the order of the flights would be dictated by weather reports as they were received, which he would review with Malmgren to determine flight destinations and routes.

Nobile's rationale for the flights was multifaceted. He viewed the 1926 *Norge* flight from Kings Bay to Teller as having been primarily preparatory. They had made history, yes, but the seventy-plus-hour flight had also proved the potential of the airship for so much more. That flight, argued Nobile, was "merely the beginning of a new era in polar exploration."

The lands and polar seas he intended to fly over had been "discovered"—insofar as they were known to exist—and portions of them mapped and charted, but mostly by men aboard ships or on the ice toiling behind sleds and dog teams, struggling just to stay alive. The regions had hardly been sufficiently explored, and certainly not from above, airborne. The unique aerial view provided perspective, relief, and had the advantage of covering vastly greater distances in the relative comfort of an enclosed cabin. Nobile would have specialized scientists taking readings, recording distances, and augmenting existing maps, which in some cases were only approximations. An additional planned outcome of the flights would be atmospheric data collected from the far north, where much of the world's weather originates, that would allow meteorologists to make better weather forecasts, potentially saving lives in the burgeoning aerial age that was on the cusp of achieving viable commercial aviation and passenger travel.

And certainly, embedded somewhere beneath all of these stated reasons for the flights, there was the reality that Umberto Nobile wanted to do it for himself and for Italy, revealing a nationalistic hubris to prove that it could be done without Roald Amundsen and the Norwegians. Nobile had said as much before leaving, in his address to the Italian Geographical Society in Milan: "We have shown how this undertaking can be executed. We prepared it, and we Italians assumed the entire responsibility for it. Why, then, should we not also reap the fruits of the work? Why not we ourselves run the whole risk and achieve the final and complete exploration of the polar regions?"

On the early morning of May 11, 1928, Nobile readied everyone for the first exploratory voyage to Severnaya Zemlya, a 14,000-square-mile archipelago lying north of central Russia. Nobile had planned a round-trip route taking them a distance of more than two thousand miles, in approximately seventy hours, without refueling. They took off without incident, but shortly after departing Kings Bay, as they cruised north along the coast of northwestern Spitsbergen, rigger Alessandrini reported that the wire controlling the rudder was badly frayed. Nobile checked it himself and confirmed the problem, determining that "the wire might snap at any moment under the strain," severely diminishing their ability to steer the airship. He was livid that this had gone undetected during repairs and maintenance, so he ordered that it be reinforced with extra wire, and they flew on. Within hours a powerful storm descended on the region, enveloping them in snow and fog. "Ice began to form all over the ship," wrote Nobile, "and the snow sticking to it accumulated in considerable quantities." Nobile well remembered the icing events that nearly brought the *Norge* down more than once, and he monitored the situation carefully.

The weather worsened, and when the radio operators gave Nobile the current forecasts, he learned that conditions for his entire route were unfavorable. Given that, and the frayed rudder cable, Nobile, after consulting with Malmgren, decided to return to base at Kings Bay and wait out the storm. Nobile's first flight from base had lasted just eight hours.

Nobile's caution proved prudent. A full-scale blizzard pummeled Spitsbergen so severely that the skies darkened despite the twenty-four-hour daylight of the midnight sun, creating an eerie atmosphere throughout the mining settlement. Heavy snow fell so hard that it began to accumulate on top of the *Italia*'s envelope in the roofless hangar. As the snow amassed, the weight bore down, and Nobile worried that the great envelope, still inflated, would be unable to bear the weight. "The stern," he wrote in his journal, "with the two great lateral fins, was so heavily loaded that the metal plating underneath began to buckle, as if about to give way." Nobile immediately sent teams of men up onto the top of *Italia*'s immense outer envelope to work in shifts shoveling and sweeping tons of snow off, and by morning they had succeeded, averting the crisis. Their boots and shovels had ripped some holes in the fabric, and these Nobile had patched with three-ply rubberized material.

After twenty-four hours of intensity, the storm subsided and the sun shone brilliantly, drying out the drooping, sodden airship. Many of the *Italia* crew were sleeping near the airship in the hangar, some curled in sleeping bags on tarpaulins, others on pallets or crates wrapped in fur clothing and blankets; they remained nearby and ready whenever called for duty. A couple of men, having just finished loading *Italia* with eighty-five flying hours' worth of fuel, sat on hydrogen gas cylinders humming fascist songs and the Italian national anthem. Outside, the skies were clear, the mountains encircling Kings Bay covered with so much snow that every direction revealed a piercing white glare punctuated by an infinite sapphire sky. Sometime midmorning, Nobile arrived and told the men they would try again for Severnaya Zemlya within a few hours.

Just after lunch on May 15, Nobile arrived at the hangar dressed in his flight suit and boots. Alongside him was Father Giuseppe Gianfranceschi, the representative sent by the pope to bless each flight. *Italia,* held down by more than one hundred men, hovered just above the ground. As the crew removed their hats and bowed their heads, the priest, according to a Norwegian journalist present, "said the explorer's prayer. The melodious voice of the Padre imparted a peculiar tone to this unusual scene, to this ceremony of departure under the vault of heaven, in nature's own cathedral, the highest dome in the world." After the prayer, a bottle of Italian spumante was dashed against the front of the control car as the airship swayed and nodded in a light breeze. Nobile posed for a few pictures, embraced his brother Amedeo in farewell, then, climbing into the control car, gave the order, "Let her go!"

The ground crew let the hold ropes ease through their palms as the *Italia* rose vertically to cheers and shouts. The engines were started, droning out the applause, and the enormous bulk of the airship soared away up the Kings Fjord and disappeared beyond the northern cape.

Nobile took advantage of the sublime flying conditions by photographing parts of Spitsbergen's northwest coast for the first time, images that would be used by future cartographers. After a few hours they reached the polar ice, and Nobile admired its character: "I looked out of the porthole," he recorded, "to see the pack, and once more I was struck by the glorious harmony between the pure white of the snow and the delicate, cloudy pearl-grey of the freshly frozen pools, bordered with blue."

The *Italia* transmitted messages to the *Città di Milano* once an hour, providing their location coordinates, elevation, and the time. These were important in the event of trouble and the need for potential rescue. Also, the reports gave the few journalists at Kings Bay at least some information to write about. In the early hours of May 16, Nobile sent the following wireless message to Benito Mussolini:

My promise to guide the *Italia* across unexplored polar territories is now being fulfilled. I send your Excellency respectful greetings from myself and my whole crew. We are now passing the northeastern point of Franz Josef Land, our course being due north. The ocean is ice free here. Perfect visibility increases my hope of landing on Nicholas II Land (Severnaya Zemlya), probably in the course of a day.

It was with tremendous pride that Nobile, flying in an Italian-built airship over one of the northernmost archipelagos in the world, using the most advanced wireless technology available—invented and built by Italian Guglielmo Marconi—could communicate with his prime minister in near real-time, by means of radio-based wireless telegraphy.

Nobile's was interested in Severnaya Zemlya because it had only been discovered fifteen years prior, by Russian explorer Boris Vilkitsky. Lying just north of central Russia's Taymyr Peninsula, the four islands of the chain separate the Kara Sea to the West and the Laptev Sea to the east. Nobile hoped to land men there for exploration and study, which would be of significant geographic and cartographic value. After thirty-four hours of flying, Nobile saw land: "On the horizon," he recorded in his log, "there appeared a vague outline, like some fantastic city of white and blue crystal rising from the ice." It turned out to be an optical illusion, and they flew on into a considerable side wind and worsening visibility.

Checking his bearings and consulting with Malmgren, Nobile determined that Severnaya Zemlya must be only one or two hours to the north. But the winds were driving them relentlessly south, and ice had begun to form on the envelope and on the outside of the control car and keel, adding weight to the *Italia* and diminishing its performance. Compounding the problems of side winds and icing, cyclonic conditions were forming north of them, making it difficult to maintain their intended course. "The squalls," wrote Nobile, "produced violent swerves of 10 to 20 degrees, which made our direction uncertain." Nobile again consulted with Malmgren. The outward flight, owing to the strong sidewinds, had taken ten hours longer than estimated and consumed a great deal more fuel than planned.

Nobile carefully considered the situation. They were now tantalizingly close to Severnaya Zemlya, but even if they reached it, current weather conditions made landing men there improbable if not impossible. So, he made the difficult decision to give up his goal and turn around, riding the wind "like a sailing ship" southwest, toward Novaya Zemlya. It was hard to relinquish his main objective, but the lives of the crew and safety of the airship were paramount, and Nobile did not feel comfortable with the current risks. Malmgren agreed, and they sped south aided by a strong tailwind.

Their return route allowed them to observe, photograph, and map four hundred miles of the Novaya Zemlya coastline never before viewed from the air. Nobile knew well the story of Dutch navigator William Barents (for whom the Barents Sea is named) overwintering on the island's northeast coast in 1596–97, a terrible tale of privation, scurvy, polar bear attacks, and death. It was exhilarating to see this harsh, dramatic landscape up close, sometimes flying just a few hundred feet above the ground.

After an uneventful four-hundred-mile crossing of the frozen Barents Sea, they reached Svalbard's North East Land on the early morning of May 18. Conditions were clear, with superb visibility, so Nobile decided to fly directly across North East Land and Spitsbergen, the deep interiors of which were also mostly unexplored. They rose to as high as 8,500-feet, and below they could see the mountains, "stretched out as on a relief map, fantastically shaped and all gleaming with ice and snow." They returned to the mooring mast at Kings Bay on the morning of May 18, after sixty-nine hours of continuous flight.

Nobile stepped down onto the snow-covered ground at Kings Bay, tired but content. Titina yipped and spun circles in the snow. Everyone was in good spirits, but also relieved to be safely back on the ground.

"It is very good to feel firm ground under one's feet again," Malmgren remarked to a reporter. Nobile posed for photographs before boarding the *Città di Milano* to consult with Captain Romagna and file reports to Rome. Although conditions had not permitted them to explore Severnaya Zemlya, they had flown 2,500 miles over 18,000 square miles of sea ice and land, making significant scientific observations. Malmgren had collected much meteorological data and had made many observations about the character of the sea ice, and the physicist Professor Pontremoli had made numerous notations of magnetic fields and measurements of the electrical conductibility of the atmosphere and atmospheric radioactivity. In these respects, and from the standpoint of aerial Arctic navigation, the flight was a resounding success. Nobile and his crew had navigated primarily by radio across vast tracts of polar ice devoid of landmarks, an altogether extraordinary achievement.

And perhaps most important of all, at least to Nobile, he had made his first protracted polar flight—one lasting three days and as long as the flight of the *Norge*—without Roald Amundsen. Now it was time to prepare for the next flight, the one the press was most interested in.

"When he goes to the North polar regions [in a few days] in the dirigible *Italia*," one journalist wrote with a mixture of doubt and wonder—for nothing so audacious and dangerous had ever been attempted before, "Nobile plans to land a detachment at the pole itself."

28

"NOTHING CAN HAPPEN TO IT."

UMBERTO NOBILE LAY IN BED, PROPPED UP WITH PILLOWS, TITINA curled in his lap. He was so racked with fatigue from the strains of navigating the recent sixty-nine-hour flight that he had agreed to an interview with a Norwegian reporter only if it could be conducted from bed. Nobile had given everyone a day off, but he was already planning the North Pole flight, via northern Greenland, taking a different, more circuitous route than he had with Amundsen and Ellsworth in the *Norge*. He wanted to depart as soon as possible, and was anxious to get back in the air, knowing that each passing day brought them closer to the prevailing summer mists and fog, and the rising temperatures which reduced the *Italia*'s lifting capacity. He was visibly drained, anxious and fidgety as the reporter asked his questions.

The reporter wished to know about the upcoming flight. Nobile responded, with a typical dose of hubris, "This [does] not present any special difficulty. After our journey to [Severnaya Zemlya], to reach the Pole seem[s] child's play." The reporter then asked about the *Italia*'s capabilities. Nobile mused for a moment, thinking of the thousands of miles his airship had already flown, through every imaginable condition. He said he believed that the *Italia* could remain in the air for at least a week, if necessary. "Practically speaking," Nobile added finally, "nothing can happen to it. Everything possible has been done to strengthen it. She is as strong as an airship can be." And with that, he asked the reporter to allow him to get more rest.

Compared to two years before, when Nobile had been at Kings Bay with Roald Amundsen, the press presence was minimal: initially just a handful of journalists on site, two of whom Nobile had brought himself. Amundsen was so famous that a circus-like atmosphere followed him wherever he went, and there had been ten times as many journalists at Kings Bay for the *Norge* flight in 1926. The diminished interest wounded Nobile's considerable pride, but he tried to focus his attentions and emotions on the success of his plans. Because his upcoming flight involved the North Pole, however, a few more journalists and film crews had arrived during the last week, with reporters and motion-picture operators from the United States, Germany, Norway, and of course, Italy.

Since the *Norge* had already flown over the North Pole, most journalists were fascinated by what Nobile called "the new and interesting . . . descent onto the ice," where men would actually be lowered onto the polar surface from the *Italia,* a maneuver he intended to perform at the pole. Taking a break from final preparations at the hangar, Nobile described the tricky and dangerous process they would attempt, which they had tested in Italy but not in the Arctic:

> The object was to carry out some oceanographical observations in which Malmgren and I were particularly interested. Above all, to sound the sea and take samples of water at various depths, down to two thousand meters, at the same time measuring temperature . . . To take these measurements we should have to anchor on the ice, or preferably in the water . . . We would bring ourselves down to within 50 meters of sea level, and from this height let down the men and instruments in a pneumatic basket (with an inflated raft which would float and could be paddled to the nearest ice mass). I myself would go down first, not only to get an idea of any possible difficulties or risks, but also to plant our flag on the ice. Once back on board I should send down Pontremoli, Malmgren, and Mariano.

For its time, the unprecedented plan was akin to the first lunar land-ing, and everything would need to be perfect to pull it off. A strong gust during the lowering of the men could prove disastrous, either killing or stranding men on the ice, but it was just such potential peril that fueled the news stories. Malmgren tried to make light of the dangers, even making an inappropriate joke about it. The Czechoslovakian scholar and scientist Francis (Franz) Behounek was a very large man, tall and fleshy. Referring to Behounek's girth, Malmgren jested, "There's an advantage for the rest of us . . . if we find ourselves stranded on the ice, about to perish from hunger, we'll always have something to live on . . ." There was uncomfort-able laughter, for everyone remembered the horrors of the Franklin and Greely expeditions, during which cannibalism had occurred.

By the morning of May 22, Nobile had checked and double-checked the airship and survival gear, and everything appeared in excellent condi-tion. The weather, which had kept them grounded for three days, improved dramatically, and both Nobile and Malmgren read Tromsø's detailed fore-cast: "The favorable situation is not likely to last much longer, because the warm currents will give rise to formations of fog. The situation is, however, still sufficiently favorable for you to start . . . provided local weather con-ditions are satisfactory."

Nobile told everyone to be ready to leave before dawn the following day. He remained up most of the night, scurrying around and directing proceedings. He confirmed that they had emergency provisions for three months, as well as smoke signals, sealed boxes of matches, pick axes, a Colt revolver and one hundred rounds, Primus stoves, and as added insur-ance, a small emergency wireless radio which he had tested and personally placed in the control cabin. He then supervised the airship inflation, and upon hearing a slight hissing sound, discovered a tiny slit at the top of the envelope, which was repaired. After further investigation by technicians, during which the *Italia* was filled with "as high a pressure of gas as possi-ble," Nobile deemed the dirigible airworthy and ordered it brought from the hangar around 3:00 A.M. By then, the high-pressure system had de-teriorated slightly, prompting the always chipper Malmgren to joke, "The weather is not the best, but one hasn't the face to keep walking the decks of the *Città* when one has already donned his flying suit!"

Once again, Father Gianfranceschi was there to say a brief prayer, and this time, he also blessed the cross which Pope Pius XI had made to be dropped at the pole. With the engines thrumming at low speed, one hundred fifty men held tightly to the mooring ropes as the *Italia* lurched and tugged just above the ground. Nobile waved to the journalists and men from the *Città di Milano* who had come to watch them leave, and when he at last stepped into the control cabin, all sixteen men, and Titina, were aboard and at their places: Lieutenant Commander Viglieri; the three scientists Pontremoli, Malmgren, and Behounek; the two senior navigating officers Mariano and Zappi; the engineer Trojani; the chief technician Cecioni; the motor mechanics Arduino, Caratti, Ciocca, and Pomella; the foreman rigger Alessandrini; the wireless operator Biagi; and the journalist Lago.

Nobile called out "Let go!" and the *Italia* was released into the sky. The men in the engine gondolas accelerated simultaneously, and the airship rose into the dawn light. A flock of gulls shrieked and scattered as the airship lofted above the shoreline and sped away. One reporter, who had been awake most of the night awaiting the moment, described the *Italia* as "a shimmering silver shape, glinting far above the ice hummocks and dimly seen from below." It receded into the sky until it blended with pewter-gray clouds forming above the horizon, melding into the swirling vapors.

And then it was gone.

29

TRIUMPH

THE EDGE OF THE PACK ICE WAS SHROUDED IN FOG. RIGGER Alessandrini traversed the ship's internal catwalks, making certain everything was operating properly, and he pulled in the handling ropes suspended below to prevent excess drag and to keep them ice free. For the first few hours, Nobile flew below the murky air, but after a time he ascended and found it diffused above 1,500 feet, so he remained there in light haze. Setting a northwest course for Cape Bridgman, at the northern tip of Greenland, they flew on into a brisk headwind. After twelve hours, soaring above ice pack cut by runnels of water leads, Nobile saw the mountains of Greenland's northern coast. Determining they had reached Cape Bridgman, Nobile noted their geographical location and changed course, steering northwest for the North Pole.

With the new course, they now were being pushed by a strong wind astern, increasing their speed, and the skies opened up into brilliant blue, the sun casting sweeping shafts of light through the cabin windows. Mariano, Viglieri, and Zappi traded turns at the rudder wheel, alternately steering the ship or measuring speed, wind drift, or taking sun sightings by sextant. Malmgren and Nobile consulted charts and discussed the weather, while Malmgren made notations on a wall chart, jotting down meteorological information received from Tromsø by wireless operator Biagi. Cecioni and Trojani took turns controlling the elevators, which altered the airship's pitch by pointing the nose up or down. The scientists

Behounek and Pontremoli read and assessed their instruments and gauges with great interest. "All on board were happy," recorded Nobile, "and contentment shone on every face."

Nobile communicated with the mechanics in their engine gondolas by telegraph, and by consulting with his chief motor engineer, Arduino, who moved along the gangway monitoring fuel consumption and directing the mechanics. Pomella manned the stern engine car, Ciocca the right or starboard side, and Caratti left or port side. With the hefty tailwind, Nobile called for running only two engines, and he gazed out the front windows of the control car. In the glorious clear skies, he could see for sixty miles, and he scanned the panorama for any unknown landmasses, but saw nothing but ever-shifting ice. Still, he beamed, his heart thrilling to the joy of Arctic flight.

Nobile had good reason to be elated. Since leaving north Greenland, they were traveling low, between five hundred to seven hundred feet over previously unexplored territory. Robert Peary's over-ice route lay far to their left, west of Greenland. The route Nobile had taken on the *Norge* was more than four hundred miles to the east, on his right. Everything below was new and never seen before.

But lurking just below the surface of Nobile's excitement was the problem of the wind. At present it was pushing them from behind at high speed toward the pole, but once they reached it in a few hours, what then? If the wind persisted, two things became clear: they would be unable to land men at the North Pole; and they would face a brutal headwind trying to return to Kings Bay. Fighting a headwind was hard on the airship as well as the helmsmen, who had to constantly correct course and zigzag as the ship's nose yawed. Nobile went to Malmgren, who stood at his wall chart, pressing his spectacles up onto the bridge of his nose. He was drawing the outlines of two nascent cyclones, one above the Barents Sea and the other above the Arctic Ocean northeast of the Siberian coast. They discussed which route might be best to follow after leaving the North Pole.

Nobile and Malmgren talked out their options. Nobile said that if the tailwind continued, it might make sense to ride it all the way south across the Arctic Ocean to the mouth of Canada's Mackenzie River, which he had dreamed of exploring. Forecasts there were for calmer weather. He had brought along navigation charts and maps of the region for just such an eventuality. The downsides of this option were that there was no organized landing prepared, and they would likely have to attempt a risky landing such as the one they had managed at Teller, Alaska, two years before. Even if they were successful, they would have to give up another attempt at Severnaya Zemlya, a vital part of their original itinerary, one they still might have time for if they returned to base at Kings Bay. Nobile asked Malmgren what his thoughts were.

"It would be better to return to Kings Bay," Malmgren said after a few moments. "Then we will be able to complete our program of scientific research."

Nobile ran his hand over his forehead. "The notion of sailing against a wind as strong as the one at present does not appeal to me," he said, "especially as I fear it might be accompanied by fog."

"No, the wind will not last long," Malmgren assured Nobile. "When we are on our way back the wind will drop after a few hours and be succeeded by northwest winds."

Nobile hoped Malmgren was right, but there was no way to be certain. The Arctic winds were fickle and had surprised them before. Nobile pondered for a time, weighing their choices. Then he had an idea that might just solve their dilemma. "Suppose," he said to Malmgren, "we were to follow the wind and utilize the favorable currents of the two cyclones to return home? The wind would carry us once more to [Severnaya Zemlya], and this time, perhaps we should be more fortunate." Weather permitting, they could possibly even land men there, finish out their scientific program, and return to Kings Bay via a route similar to their previous sixty-nine-hour flight.

It was a tantalizing prospect, but after studying the charts once more, Malmgren did not like the look of the atmospheric conditions around Severnaya Zemlya. He remained convinced that returning to Spitsbergen directly from the North Pole was best, and Nobile at last conceded that he was probably right.

Around 10 P.M. that night, after eighteen hours of flight, the weather changed. Before them, rising from the horizon perhaps an hour away, loomed a menacing cloud barrier over three thousand feet high. "With its weird outlines," Nobile wrote, "it looked like the walls of some gigantic fortress." Nobile now worried that they might not even make it to the North Pole, and that the press around the world—and Mussolini and the Italian people—would excoriate him for failing. They flew toward it, rising above darkening clouds that guarded their destination like battlements, to obtain a sun shot to determine location. They were at 88° 10′— just fifty miles, and less than an hour, from the pole. Providentially, the dark cloud mass thinned and dispersed as they careened on at seventy miles per hour, with the tailwind so forceful it sometimes took two men each at the rudder and elevator controls to fight the bucking and maintain the airship's trim. At 12:20 A.M. on May 24, one of the officers looked up from his sextant and cried out, "We are there!"

Nobile called for the engines to be slowed and ordered the helmsman to descend through the residual haze. A few minutes later, they could clearly see the pack ice below, buckled and heaving with the constant churning and grinding of floes. "All engines, half speed," Nobile ordered Chief Engineer Arduino. "Helm full circle, elevators down." As they circled, battered by a southeast wind each time they came around broadside to it, Nobile readied for the ceremony he had promised his pope, his country, and the city of Milan. The journalist Ugo Lago scribbled furiously in his notebook, writing messages for Biagi to send when he could.

Despite having been there two years before, Nobile was overtaken by emotion. This time was different. This was *his* triumph, not Amundsen's. As expedition and airship commander, he alone would reap the glories. He intended to share the accolades with his handpicked crew, but everyone knew that the largest wreath of laurels always went to the man in charge.

At 450 feet above sea level, Nobile leaned from the cabin door and committed the Italian tricolor flag to the wind. Next, he tossed the flag of the city of Milan, embroidered with a coat of arms; it sailed and then spun, its wooden staff twirling like a spindle. Finally, with Cecioni's help because it was so heavy, Nobile pushed from the open door the pope's hand-hewn oak cross, wrapped with cloth bearing the green, white, and red colors of Italy, and it fell onto the ice not far from the other flags. The officers stood solemn and proud, saluting during the short ceremony.

They closed the doors, and with the engines at their lowest speed and the propellers turning slowly, the inside of the control cabin fell quiet. To Nobile's surprise and delight, Lago put a record on the gramophone, and as they circled the flags and cross atop the world one last time, the sounds of the well-known folk song "The Bells of San Giusto" rang through the cabin, transporting the Italian contingent momentarily away from the tempestuous air above the ice and back to their homes, their families, to Italy. "We were all moved," Nobile wrote of the moment. "More than one had tears in his eyes." Then they all drank a toast of warm liqueur, milk, egg whites, and deep, rich wine.

Finn Malmgren, though not Italian, was wiping his eyes too. He clasped Nobile's hand warmly and cleared his throat. "Few men can say, as we can, that we have been twice to the Pole."

"Few men indeed." Nobile smiled. "Six Italians and one Swede."

After two hours of circling the Pole, as Biagi sent coded messages of their success to the pope, the king, and Mussolini, Nobile made the final decision. The winds were too strong to risk lowering men onto the ice. He gave the order, and the *Italia* banked around. They were headed back to base at Kings Bay.

30
"IT'S ALL OVER!"

THE AIRSHIP *ITALIA* PLOWED INTO A MORASS OF HEADWIND, FOG, and snow, making little headway. Wind sieved through cracks and joints in the cabin in a continuous, high-pitched shriek, sounding at times like piercing screams. Ice shards flung from the propellers struck the envelope's flanks, lashing at the rubberized cotton fabric. For the last twenty-four hours they had churned through the gale, and by their best estimate— only approximate since they had not seen the sun since leaving the pole and were flying solely by magnetic compass—they were only halfway to Kings Bay. The southwester blew them off course eastward by twenty to thirty degrees, with no sunlight through the fog above them—and below them only endless monochromatic ice pack—they didn't know where they were. Everyone was frayed to near delirium.

Nobile had been awake for nearly three days, and though he would not admit fatigue, his eyes shut periodically, and he would nod off for a moment, then shake himself back to consciousness. He had to stay awake and guide the *Italia* to safety, but their circumstances were dire. Some time ago they had dipped down below the fog to six hundred feet and, taking measurements, determined their ground speed—even with the engines running at full tilt—to be just twenty-five miles per hour. They were barely moving, and definitely not in a straight line. At 3:00 A.M. on May 25, Nobile recorded that he was "Worried by the high petrol consumption and the strain . . . on the structure of the ship." Indeed, for the last few hours, everyone in the control car had felt the entire internal framework of the *Italia* quivering and straining as the massive airship lurched and shuddered through the turbulence.

Everyone worked at their stations in the silence of their own worries, all racked with exhaustion. Malmgren took long turns at the rudder and elevator wheels, giving the others a rest. Malmgren felt terrible that his prediction of wind shift had failed to materialize, so he worked harder than ever before. Nobile continued to move, now zombie-like, between the navigation table, the wireless cabin, and the speed instruments. At one point, up in the front of the pilot's cabin, he paused at the photograph of his daughter Maria he had tacked to the wall. He had brought the same picture along on the *Norge* for good luck, and as a reminder of his life back home. Seeing it now, in his sleep-deprived state, her eyes seemed to come alive, staring back at him. "I was struck by the sadness of their expression," he wrote, "they seemed to be misted with tears." He may well have been projecting his own emotional condition.

At 7:30 A.M., Biagi radioed the *Città di Milano* at Kings Bay, giving them a rough estimate of their position through dead reckoning, but because their last known fixed point had been taken six hours prior, they might be off by one hundred miles or even more; it was impossible to tell. Biagi had been receiving intermittent reports from the base ship, including weather and compass directions of dubious accuracy. He consulted with Nobile, who reported that he had been anxiously expecting the Spitsbergen coast to come into view at any time, but every time he looked out the windows, there was nothing but fog and ice all around them, and a thickening skin of ice was forming over much of the envelope. Biagi sent a quick message to base: "If I don't answer, I have good reason."

Two hours later, Nobile was standing by the door of the wireless cabin, hoping for any news at all from Kings Bay that might help Biagi get a fix on their position, when he heard Trojani shout, "The elevator wheel has jammed, nose down!" Nobile rushed to his side and tried to help him move the wheel, but it was jammed. They were at 750 feet, with the elevator fins forcing the dirigible downward. Nobile knew that if they kept on like this, within a few minutes they would strike the ice pack.

"All engines, emergency stop," he commanded, his voice now hoarse and thin. For a few tense moments, everyone stood still, waiting for an outcome. At 250 feet above the ice, with no engines propelling them forward and lightened by fifty-four hours of fuel consumption, the *Italia* began to rise gently. Still, they needed to steer and control the airship, and Viglieri came to assist Trojani. Viglieri used brute force and with a blunt hammer blow, jarred the elevator wheel loose. Nobile called Cecioni to examine and repair the elevator wheel, which appeared to have jammed due to icing. He also instructed him, as a precaution, to untether and prepare to deploy the heavy ballast chain and ice anchor, which could buffer the impact of a forced landing. One similar to it had worked well enough at Teller, helping them safely land the *Norge*.

As they continued to ascend, Nobile checked the pressure gauges and noted that the pressure in the last stern compartment was higher than the others, so he released enough hydrogen to equalize them. Just then, Mariano approached Nobile and asked if they could go ahead and continue rising out of the fog, where he might take a sextant sun reading to try to determine their location. It was a difficult decision. The fog did seem to be dispersing, and he could see sunlight above. Taking a sextant shot would be helpful but depending on the altitude required to get the reading, there was the very real risk of the increased temperature of the sun causing rapid expansion of the hydrogen gases in the ballonets, with potentially disastrous consequences. And he could ill-afford to valve off too much hydrogen as a response, for that would make the airship heavy and reduce crucial lifting power needed to remain airborne. It was, as always, a delicate and imperfect balancing act. After a few minutes of silent deliberation, Nobile deemed the risk worth it, and upward they sailed, still with no engines running, cutting through the mist and reaching luminous, blue sky at three thousand feet.

Mariano got his sextant shots, and they remained above the band of surface fog for nearly a half hour, rising as high as 3,300 feet. Apparently oblivious of the time in the warmer temperatures of the direct sunlight, Nobile scanned in all directions with field glasses, looking for some sign, any sign of Spitsbergen's high peaks, but all he saw was sky and a dense layer of murky air hovering over the pack. At that point, Nobile again checked the pressure gauges and realized that the hydrogen gas had indeed begun to dilate; he quickly ordered the port and stern engines started and called for a descent back down to near the pack ice to determine their ground speed and drift.

Back down at nine hundred feet the ice came clearly into view again. With just the two engines operating, they were now flying at about thirty miles per hour, indicating that the headwind had died down. Malmgren took little solace in this, as it had finally eased off more than a day after he had predicted it would. Nobile spoke with Mariano and Viglieri, and the three of them determined that, as best they could tell using all of the available recent readings, they were approximately 180 miles northeast of Kings Bay. Nobile was relieved. In a little while, after everyone had determined that the airship was running properly again, Biagi could radio that they expected to be back by three or four in the afternoon.

Nobile looked around at his stalwart but stressed crew, thankful that their most recent trials seemed to be behind them. Perhaps, in a few hours, he could get some much needed rest—he had now been awake for seventy hours. Everyone was at his post: Malmgren manned the helm, with Commander Zappi standing beside him, offering counsel. Cecioni remained at the elevator wheel, which was now operating smoothly; his hands were covered with grease from repairing the mechanism. Next to him, watching the pressure gauges and maintaining communication with the engine mechanics in their gondolas, stood Trojani. General Nobile sat at the navigation table in the rear of the control cabin with Viglieri and Mariano, who were working together plotting distance and speed coordinates on the chart. Behind them, near the compass, Professor Behounek busily worked with his instruments.

Foreman Rigger Alessandrini moved along the "balloon deck" gangway, inspecting the internal framework, the state of the envelope, and checking the gas cells and valves for any leaking hydrogen. Above, Arduino made his way along the steel catwalk leading to the three engines, peering outside through porthole windows to ensure they were running properly and that the propellers were not icing or failing. In a cramped bunk area near the stern, Professor Pontremoli and journalist Lago were sound asleep. The engine mechanics were outside the airship in their respective engine gondolas: Pomella in the stern gondola; Ciocca in the right and Caratti in the left.

The *Italia* was cruising along at an altitude between six hundred and nine hundred feet. Nobile rose from the navigation table and moved forward to the front of the control cabin and stared out the right porthole, mesmerized by the rough and ruptured ice below. No matter how many times he had seen the pack surface, it was always hypnotic, the blues and whites and sometimes emerald greens where light adorned it, the ice sheet always moving, a living thing. To cross-check their altitude against the altimeters, Nobile reached out and dropped a glass ball filled with red dye, marking the time on his stop watch until it struck the ice and exploded, stippling the ice red as it hit. Ten seconds . . . 710 feet.

He was about to drop another ball when he felt a shift under his feet, a sudden lurch as the airship listed toward the stern. As he adjusted his footing, Cecioni, at the elevator, called out with alarm: "We are heavy!"

Nobile quickly consulted the instruments. The inclinometer showed them out of equilibrium, with the stern down 8 degrees, the nose pointing upward slightly, and they were losing altitude, dropping tail-down from the sky toward the ice pack at a rate of two feet per second.

"All engines. Emergency. Ahead full power!"

Nobile watched the instruments, trying to will or conjure some lift, but even at full power they continued to plummet at twenty to thirty miles per hour with the nose tilted up at 20 degrees.

"Up elevators," he shouted over the wail of wind. Alessandrini arrived, his face pallid with worry, to see what was wrong. "Hurry and check the stern valves," Nobile barked at him, and Alessandrini disappeared through the hatch in the roof of the cabin and hustled along the gangway back toward the ship's stern to inspect for a leaking valve or holes in the envelope.

"Look, there's the pack!" Malmgren yelled from the steering wheel. Through his window he could see fractured rubble just a few hundred feet below and coming up fast. Everyone worked furiously, but they had lost control of the airship. Now Nobile could only mitigate disaster and avoid fire or explosion at impact. He sent Cecioni to heave the ballast chain, and Zappi took over at the elevator wheel.

"Stop all engines!" he ordered, a weird calm of resignation in his voice. "Close all ignitions!" Within a few seconds two of the engines had stopped, but the left one, run by Caratti, was still running.

"The wheel is dead!" Zappi shouted. "The elevators have lost all response!"

Cecioni was grappling with the ballast chain, but it was entangled with rope. Nobile felt the airship banking hard to one side, and he thrust his head out of a porthole and saw that the propeller on Caratti's engine was still whirring at full speed. Looking behind, he saw that the stern engine was just thirty feet above the ice, and spindrift and ice fragments scattered and swirled in the frozen air of the wake. He tucked back inside the cabin and fought his way forward to get between Zappi and Malmgren. Malmgren wrestled with the twisting wheel and Nobile grabbed ahold to help him, trying to somehow guide them in to a soft snowfield, but the wheel gave no response and he saw only giant ice boulders and ragged masses of ice and snow growing larger.

Malmgren threw his hands in front of his face.

Nobile closed his eyes.

A sickening jolt. A blast of shrieking wind and shattering glass. The crush of bending steel.

"It's all over," Nobile whispered to himself.

31
IMPACT

AS THE CONTROL CAR STRUCK THE ICE THE FRONT SECTION PUL-
verized, and most of the car's underside sheared away. Nobile felt a whoomph
of freezing air, heard tearing canvas, then there was a blunt bash to his head,
somersaulting him forward, and then a grotesque and unmistakable snapping of limbs. The remnants of the control car skidded across the harsh
surface ice; inside was a tornado, a whoosh-roar of metal and glass and
paper and splintered wood. They lifted for a moment, airborne until the
Italia bounced once and the rest of the control car was torn from the envelope with a violence like an explosion, and men and equipment, crates and
gear, cables and ruptured instruments were scattered across the sea ice as
the rest of the *Italia,* lightened by a few tons, lofted back into the air and
soared above them.

Commander Nobile opened his eyes. He was on his back. Cold. Quiet.
Blood ran down his forehead. Titina whined next to him, shivering.
Above, to his left, silvery and pewter gray against the slate sky, was his
dirigible. She hovered, still inflated, her underbelly lacerated as if vented
by some giant fisherman's blade. Near the gash where the control car had
been ripped away, canvas remnants flapped like fins, and spilling out, like
dangling viscera, trailed ropes and cables and the snapped ends of metal
tubes. Nobile regarded his 350-foot-long mutilated craft, floating there
fifty or sixty feet above him, a free balloon, a derelict ghost ship. He could

see that the two side engine gondolas remained attached, but the stern gondola was gone. As he tried to sit up, sharp pain seared through his right arm and leg. He clutched at his chest with his left hand, then slumped back down, gazing at the debris field left by the crash.

Malmgren squatted just a few yards away, mumbling in Swedish. Beyond him, Cecioni lay writhing in pain; then he stopped, looked up, and gawked into the sky, his arms outstretched. Seconds before he had tried to get up, to run, to do something, but one of his legs was shattered below the knee. He stared, dumbstruck at the airship hovering above, and he could recognize men still aboard: there were Lago and Pontremoli, who had been asleep, peering down through a great rent in the envelope, their faces wrought with terror. Not far from them, Alessandrini clung to a bent spar of internal framework, dumbstruck horror on his face. Beyond those three, someone else—it must be Arduino—was hurling boxes and tanks and crates from the hole where the control car had been, and these fell to the ice as the airship now rose higher, lifting faster and growing smaller until at last the bold black insignia *ITALIA* was illegible and the dirigible melded with the clouds and dissolved into the ether, the men with it.

Nobile surveyed the scene of the crash site. There was a long scar gouged into the surface where the control car had skidded, and weirdly, along the entire length of the track, a stripe of bright red. Nobile cringed at the sight, which looked "like blood which had flowed from some enormous wound," then realized it must be red paint from the exploded glass balls used to measure height. Beyond the carnage, he saw nothing but an amorphous confusion of ice extending to the far horizon.

Things started coming into focus. Malmgren now hunched near him, one arm hanging awkwardly at his side. He continued muttering, alternating between Swedish and English. Mariano was standing, looking around, saying over and over, "Va bene, va bene . . . it's all right. We are here, together" as if to convince himself that they were actually alive. Not far off, Cecioni was moaning and clutching at his legs. Zappi lay a few yards off, eyes open, in a silent stupor. Three others were upright and staggering about, battered and bruised but otherwise uninjured: Viglieri, Trojani, and Behounek. Biagi stumbled up to the group, his breathing labored. He was clutching the emergency radio to his chest. He had grabbed it right before impact. "Our field radio is intact," he husked.

Nobile, his voice raspy, looked at those assembled as wind blew across the ice plain. "Lift your thoughts to God," he said, though his chest hurt when he spoke. After a few moments of silence, his voice strained nearly to tears, he added, "*Viva l'Italia, Viva l'Italia*," and the others joined in with a unified, mournful chant: "*Viva l'Italia.*"

Zappi, trained in first aid, went around checking everyone's injuries. Nobile's right arm and left leg were broken, and his chest pain indicated fractured ribs. The blood on his head had dried, and although he likely had a concussion, he was lucid and giving orders, telling others to find anything useful on the ice among the dim outlines of bags and crates strewn in the distance. Cecioni also had a broken leg, and Mariano and the others managed to prop Cecioni next to Nobile where they could be cared for together. Trojani trotted back with one of the emergency sacks slung over his shoulder. He knelt down and, his fingers trembling with cold, untied the nylon cording at the sack's neck. He pulled items from the large bag: boxes of pemmican, sealed packages of chocolate, a single fur-lined sleeping bag, a tent, a Colt revolver and one hundred cartridges, matches, and a signal flare. This was good, it was something. They covered Nobile and Cecioni with the sleeping bag and began setting up the tent. Unfortunately, no medications or medical items of any kind had been located. Among the debris blowing around on the ice, they found pictures that had adorned the walls of the control car: an image of *Madonna of Loreto,* one of Mussolini and one of the king, and Nobile's picture of his daughter, Maria. Nobile kept Maria's picture in his vest pocket, and they displayed the Madonna by securing it to the center tent pole.

While the less injured scoured the crash site for anything else they might use, Malmgren remained near Nobile, crouched and holding his shoulder, which felt to him either broken or dislocated. Malmgren's face was swollen and bruised, and he stared off into the sky, as if still looking for the *Italia*. His blue eyes were glazed over, and he stammered, in a state of shock.

Nobile tried calm him down. "Nothing to be done, dear Malmgren," he said, but before he could continue, Malmgren replied, "Nothing but die . . ."

Malmgren stood slowly, hunched over, cradling his wrenched shoulder with his good arm. "General," he said, looking out into the distance, "I thank you for the trip. I go under the water . . ." And he turned and walked away. Nobile watched him take a few unsteady steps, confused by what he had said. Then it occurred to him: he was going to drown himself in an open lead somewhere.

"Malmgren, No!" Nobile yelled. "You have no right to do this!"

Malmgren paused and looked back, perplexed. But he was listening.

"We must wait. We will die when God has decided," Nobile said to him.

For a few moments, Malmgren stood motionless, considering. Then he walked back and, wincing with pain, sat down on the ice next to his commander again.

Biagi roamed around searching for any metal parts or pieces of steel tubing he might use to construct an aerial for the radio. He followed along the stripe of red paint, seeing all manner of wreckage strewn on either side of the macabre-looking red band. Snapped off control cables lay strung out like snow snakes, and tangled guy wire was wrapped weirdly around an ice chunk. Then Biagi noticed something in the distance that looked like an ice hummock, but as he drew closer, he could see it was part of the airship, the remnant wreckage of the stern engine car. Next to the demolished engine car, sitting on an ice block, was Motor Mechanic Pomella. One of his boots was off, and he was hunched forward as if to put it back on.

Biagi approached him. "Are you all right, Pomella?" When there was no response, Biagi reached out and shook his shoulder to rouse him, and

Pomella tumbled forward and rolled onto his back on the ice. A dark stain of dried blood surrounded his mouth and had frozen like a beard at his chin, and one side of his face was purpled by bruising and stove in at the cheekbone.

Vincenzo Pomella, just thirty years old, was dead.

32

"SOS *ITALIA* . . . SOS *ITALIA*"

NOBILE LAY SHIVERING ON THE ICE NEXT TO CECIONI, WHO WAS half conscious, yelping periodically in pain like a dog having a nightmare. Nobile could see men nearby erecting a tent on a flat section of ice; around them were bags and piles of gear. Nobile called out for Mariano, who as a commander and first officer, was second in charge of the expedition. Mariano trudged over from the tent site.

"I feel myself dying," Nobile said to him. "I think I have only a few hours to live." It hurt to speak. With fractured ribs, pain ran through him with every breath. For some time, he had been going over the crash in his head. From the moment the elevator wheel jammed, it had taken just two minutes for them to plunge from the sky. Now, things appeared hopeless. The pain was so great Nobile thought he must have internal injuries, too, and likely internal bleeding. Coupled with a broken arm and leg, he would be unable to lead his men.

"I cannot do anything for you," he said to Mariano. "Do all you can . . . to save our men."

Mariano pulled the sleeping bag higher up on Nobile. "General, set your mind at rest. There is still hope. We have the wireless . . . and very soon we will be in communication with the *Città di Milano*. And we have picked up a case of provisions too. We can hope."

Hearing those reassuring words, Nobile asked for a head count and

an inventory of everything available to them. Mariano reported that Pomella was dead, and that they had covered his body with parts of the
engine gondola. There were nine men on the ice still alive, including Nobile: Mariano, Behounek, Malmgren, Zappi, Trojani, Biagi, Viglieri, and
Cecioni. Mariano assured Nobile that Biagi was working hard to get the
wireless radio operating. He had found some steel tubing to fashion an
antenna and was confident it would soon be working. They had found 160
pounds of pemmican—so they had food for a time, perhaps a month on
low rations. And one of the men was working to melt surface ice for drinking water. They had one four-man tent, which was now set up; it would
be cramped, but it was shelter and they would make do. In a little while,
Mariano told Nobile, they were going to splint his and Cecioni's fractures
as their injuries were most severe.

Nobile thanked Mariano, then closed his eyes. His tortured thoughts
went to those last seen aboard the *Italia,* drifting away toward the northeast. He pictured Ettore Arduino, the skilled chief engine engineer. Selfless,
he was still throwing provisions down onto the ice as the motorless airship
floated away. Dependable, unflappable, and always positive. He thought
of Engine Operator Attilio Caratti, handsome and fit. He remembered,
after the *Norge* Expedition, how the American girls had swooned over his
elegant good looks. And dear Renato Alessandrini, who also served with
him on the *Norge*. Nobile recalled his wonderful sense of humor, and how
he had teased him for being too fat to fit through the hatch leading up into
the envelope, and how Alessandrini had only laughed.

Nobile convulsed in a shudder of cold and revulsion, thinking of them
and what might have befallen them by now. Now came the image of Calisto Ciocca, the slender mechanic who could work for days on end, always
with a smile and a song. And journalist Ugo Lago. Nobile sat upright
for an instant, stunned by the cruelty of the young reporter's fate. Back
at Kings Bay, both journalists had wanted to go, but Lago had won his
spot by a coin toss. Nobile had liked his fiery Sicilian personality, his loud
and volatile voice. And finally, there was Professor Aldo Pontremoli, the

brilliant physicist from Milan. His exertions aboard the airship had impressed Nobile deeply. When not doing calculations, he was always helping in some way; he had even assisted the riggers repairing the torn envelope.

Now, all of these men were somewhere—but what might have happened to them was unnerving to think about. They might still be aloft, blown dozens of miles away. Or, maybe, if luck was with them, they might have come down and survived, and were themselves gathering emergency equipment and food. There was much still aboard. Nobile prayed for this outcome, and he vowed, if and when they could, to send rescuers in search of them. But in the moment, all Nobile and the eight crash victims could do was try to save themselves.

Nobile awoke from fitful delirium—it might have been five minutes or five hours, he had no idea—to see Zappi kneeling next to Cecioni, telling him to calm down and remain still. Through the muted light Nobile could see he was now inside the tent, and he vaguely remembered men carrying him there as pain jolted him with their every step and stumble. Afterward, he had fallen into half consciousness. Now Zappi was wrapping Cecioni's leg with strips of canvas cloth from the control car and a splint of wood from the wreckage. Cecioni cried out in pain, but Zappi worked fast. "Now it's your turn," he said, looking over at his commander. Nobile tried to force a smile, but he knew there would be pain. Zappi managed to wriggle Nobile from the sleeping bag as his commander hissed through gritted teeth, the sickening sensation of his snapped right tibia, jutting at an unnatural angle, nearly causing him to vomit. Zappi splinted Nobile's leg and arm, then cut the sleeping bag down the side and managed to get both Nobile and Cecioni inside it together, since they could not move or walk around to produce any of their own heat.

By late afternoon on May 25, the makeshift emergency camp was well organized on the ever-shifting ice pack. The pemmican, chocolate, flares, matches, and spare radio batteries—to prolong their life by keeping them from the elements—were in one corner of the nine-by-nine-foot silk-walled tent. Nobile and Cecioni, their legs outstretched because they couldn't bend them, took up about a third of the tent's floor space. To sleep, others would have to curl and hunker atop one another and slumber as best they could. All the men gathered for a time to eat some pemmican together and discuss their situation. Nobile and Mariano agreed that at eleven ounces per man per day, they had enough pemmican for twenty-five days. They talked about the men left aboard *Italia,* and Biagi said that at one point, while working on the radio antenna, he and Zappi had seen a thin column of smoke in the direction the airship had disappeared. The airship may have crashed and exploded, or they might have survived and used petrol to light signal fires. Or, it might have been vapor, or clouds, or an Arctic illusion—Biagi and Zappi could not be sure.

Later on, Mariano slipped into the tent and handed Nobile the Colt revolver for safekeeping. In a low whisper, he confided to Nobile that he had come upon Malmgren out on the ice, one arm in a sling, holding the pistol in his good hand. Malmgren had been stammering something about being of no use, and Mariano had acted quickly, stopping Malmgren before he could harm himself. He had chastised Malmgren for even thinking about it, then softened, telling the Swede that they needed his Arctic skills if any of them were to survive. He had been with Roald Amundsen on the *Maud* Expedition—and knew the behavior of the ice, and the vagaries of Arctic weather. As Malmgren paused and turned to meet Mariano's eyes, Mariano had quickly taken the gun from his hand and led him back to the tent site.

In low tones, with rising winds outside flapping the nylon tent, Nobile made Mariano assure him he would keep a close watch on Malmgren. He was obviously despondent, and based on some of the things he had said, blamed himself for the accident, for his bad prediction about the wind direction. The headwind had not let up before it had done irreparable damage to the airship, and it was his fault. Nobile implored Mariano to convince him to the contrary and restore his morale and said he would do the same. Mariano agreed and went back outside to continue searching the crash site. They had been finding more bags of gear behind hummocks and in crevasses.

Biagi poked his head inside the tent door to tell Nobile and Mariano that he thought he had the radio working. He had sent his first test messages—with no response. The three men went over the last terrible minutes on the *Italia,* working out the time and rough coordinates in their heads. The airship had crashed at about 10:30 A.M. Greenwich Mean Time. By their collective reckoning, they guessed they were somewhere north of Moffen Island, perhaps 180 miles northeast of Kings Bay. Nobile told Biagi to keep up the same communication with the *Città di Milano* that they had established on the airship: transmit on the fifty-fifth minute of each odd hour, and then listen for a reply.

That evening, everyone crowded into the four-man tent, exhausted, still in shock. Some chewed listlessly on chunks of pemmican, speculating on the fate of the six unfortunates who had sailed off into the sky; others dosed into fitful sleep. Nobile, through searing pain, jotted recollections of the first night on the ice in a little notebook he had found in his breast pocket: "Nine men huddled together in . . . a cramped space. A tangle of human limbs. Outside, wind . . . howling."

When it was time to make the radio transmission, Biagi rose and left the tent. He shuffled across the ice to his makeshift wireless station, his aerial antenna constructed of twisted tubing from the dirigible's framework, and extended from this, "a number of lengths of bracing wire that had snapped off at impact." He checked his watch, and at precisely five minutes before the hour, began transmitting.

"SOS *ITALIA!* . . . SOS *ITALIA!* . . . SOS *ITALIA!*"

33

THE RED TENT

NOBILE AWAKENED TO EXCRUCIATING PAIN COURSING LIKE ELEC-tric shocks through the entire right side of his body, the bolts of current soon replaced by intense throbbing and aching. As he opened his eyes, he saw Biagi and Zappi kneeling on the fabric tent floor tinkering with the radio transmitter, which they worried was not sending messages. So far, although Biagi had been sending SOS messages every other hour through the night, they had received no response. With Zappi's assistance, they had dismantled the radio, cleaned all its parts, and checked for faulty connections, and were now carefully putting it back together again so they could keep trying.

Malmgren was also in the tent, slumped against a wall, but looking somewhat improved. Color had returned to his face, and his eyes darted about. "I'll see about the water," he said, sitting up and crabbing his way around the center pole toward the tent door. By now, everyone was thirsty, their throats dry. Malmgren knew how to find freshwater surface ice, but he would need to make a boiling pot of some kind to melt it.

Men returned periodically from scouring the area to report found items. A few cans of petrol had been discovered intact, vital for making fires to heat food and water, and also for smoke signals. More food had been located: meat extract, butter and sugar, malted milk—all told, combined with the pemmican, nearly three hundred pounds. Nobile figured it was enough for them to survive on for six or seven weeks. Then, someone trotted up to the tent and with excitement reported that they had found two sextants, a mercury artificial horizon, chronometers and calculating tables—along with two navigational charts of the area—as well as a copy of the *Arctic Pilot,* a nautical guide of the region's seas and coastlines. These instruments and charts had been on the table and wall of the control cabin that had torn apart at impact, and by luck all appeared to be in good shape. Nobile was pleased, for now they could report their estimated position if the radio worked. In the meantime, he would continue to pray that someone heard them.

Later in the day, the clouds parted long enough for Mariano to shoot some sun sights, and he calculated their position at 81° 14′ north latitude, and 28° 14′ longitude east. It was only a rough estimate, since the sun was not out long enough to get multiple sightings for cross reference, but they believed they were somewhere northeast of Charles XII, Foyn, and Broch Islands, three small islands directly above Svalbard's uninhabited North East Land. Measuring the drift of the current, they found they were moving toward the southeast, which might possibly carry them near one of the islands, or even twenty to thirty miles farther to the North East Land coast. Everyone was buoyed by the news. Although there would be no people there, at least they might be able to get off the ice, which was continually moving, shifting, and breaking up. As warmer summer temperatures arrived, the breakup might completely disintegrate the floe they were on and engulf them.

The tent site chosen was on flat ice that was slightly elevated, with the hope that they might be more easily spotted by search and rescue planes. They believed that by now, the *Città di Milano* should have received their SOS calls; if not, they would have realized something was wrong. In either case, they hoped someone was organizing a plan to find them. To make their tent more visible to potential rescuers, the men painted the outside walls with the bright red dye from some of the unbroken glass balls called "altitude bombs" found along the crash site, beside the flanks of the long red stripe tattooed into the ice. When they were finished smearing the silk tent walls with dye-soaked rags, they dubbed their little pyramid-shaped shelter "The Red Tent" and stored the remaining intact dye bottles away for later.

By the end of the second day on the ice, Malmgren had developed a system for melting freshwater ice. He had cleaned out an empty petrol can as best he could to use as a pot, into which he placed chunks of older surface ice to heat. Malmgren understood, from his time with Amundsen, that new ice or "young ice" had brine trapped in the crystals, making it too salty to drink. But as ice ages, the brine leaches out of it, and older, multi-year ice was drinkable once melted. Malmgren scouted around and found sections of older ice and started a daily ritual of boiling and storing fresh water for the men, which was so precious it was only distributed at mealtimes, when Nobile handed out rations of a half pound of pemmican and a few pieces of chocolate. The fresh water was so hard wrought that it was doled out in a thermos lid which was passed around from one man to the next. Malmgren, now feeling useful to the survival of the group, seemed in a much better mental state, conversing with Nobile about the weather, their drift direction, and the condition of the ice. Nobile was relieved to see his improvement after the two suicide threats.

Cecioni, however, was not faring well. At night he squirmed and thrashed in agony from the pain in his broken leg, sometimes clutching at Nobile in a fit of panic that awakened the commander. Once he struck Titina for trying to curl in next to him and Nobile had the big man moved away to the other side of the tent so he could get some sleep. Nobile recorded in his journal, "A great effort to bear the nervous distress of Cecioni, who at time seems . . . mad . . . wide-eyed with terror." Cecioni felt certain they were going to die out on the ice, either from exposure, starvation, being swallowed by the ice, or torn to death by polar bears, and the many possibilities plagued him with despair.

Biagi, satisfied with his repairs, set the radio transmitter up again and, bundled up in his lambswool flight suit, with his hood pulled down over his head, sat on an overturned fuel can dutifully sending out messages and craning his ear to the set listening for some response, any response. It was hard to hear anything over the constant wind now blowing hard from the

north, covering him and his equipment with a sheen of crystallized spindrift. Dark clouds hung overhead, casting a pearl-gray pallor across the
icescape, with mist rising in vaporous columns all around him.

Nobile knew the importance of routine, a structure of duties and responsibilities among the nine survivors. On the *Italia,* everyone had numerous jobs and were nearly always busy, unless at rest. Now, despite their
dire circumstances, he wanted those who were uninjured to remain active.
He and Cecioni were too injured for physical duties, so they would remain
in the tent and measure, plan, and dole out the scant meal rations. Each
member was designated a location in the small Red Tent, which would be
their place to rest and sleep unless they were on watch, another duty that
involved scanning the skyline for search planes or land, which Malmgren
said to watch for toward the southwest. They should also keep a lookout
for polar bears.

Malmgren, though reduced to one good arm, was tireless in his search
for suitable ice for fresh water, which they were now using also to make a
pemmican soup—more of a kind of yellowish-brown gruel—but at least it
was warm. Everyone was suffering during the freezing nights, having only
the flight gear they were wearing inside the airship when it went down.
Lost was nearly all the emergency Arctic gear Nobile had so painstakingly
selected and packaged with Fridtjof Nansen's counsel: reindeer skin boots,
snowshoes, skis, and all of the other sleeping bags. Nobile and Cecioni
were taking turns with the one fur-lined bag, and the rest in the tent
shared a single blanket, trying to spread it over four or five men at a time,
the ice below them sucking away body heat.

On the evening of their second night on the ice, Biagi called for others
to come listen to the radio with him. He told everyone to hush as they
hovered around the radio, and just above the light whistle of wind they
could hear an Italian voice, faint but audible. It was the *Città di Milano*!
"We have not heard your radio," the operator said. "We are listening for
you on the nine-hundred-meter band on shortwave. We believe you are
near the north coast of Spitsbergen. Trust in us. We are organizing help."
At first, there was a cheer of elation. They were looking for them! But

the more they listened, the more their spirits fell. The same message kept playing over and over, until they realized that those on the *Città di Milano* had not heard Biagi's messages and were unaware of their coordinates. The survivors were actually nearly two hundred miles from the north coast of Spitsbergen. It was simply a stock message, interspersed with other messages from crewmen aboard the ship to family members back in Italy. But at the very least, Nobile told the men to shore up morale, Captain Romagna and his officers had deduced that something was wrong.

Late that night, as everyone but one lookout lay shivering inside the tent, there came a percussive crack like gunfire, followed by elongated crashing noise, then an eerie screech-wail. "Everyone out of the tent!" Malmgren was yelling. "The pack is breaking up!"

Men flung open the tent door and crawled out, dragging Nobile and Cecioni behind. They propped up the invalids as ice ruptured all around, shaking the tent and radio base in an ice quake. Everyone waited and listened to see if the tent would be swallowed up. After a few tense moments, the splitting and cracking subsided, replaced by a low rumble, like the far-off sound of a distant avalanche.

Nobile gazed around at the ice pack and could see upthrust ruptures, pressure ridges forming not far from their camp. It occurred to him, studying the weird and alien surface, that the moon's landscape must be similar. Malmgren spoke up: "This is a bad position," he said flatly. "We better look for another place . . . the ice here is so broken and jagged . . . it is more exposed to the winds and currents."

Nobile agreed that some men should scout for a safer camp site. A few men fanned out, seeing small leads of open water snaking around their floe. Much more breakup and they would be stranded on a very small ice island. Then Nobile heard one of the men—he could not tell which—shouting. Nobile wiped his eyes, blinked, and strained to see what was going on. The shouts came back in a relay, and Nobile sat up straight and squinted at the horizon. Beyond the hummocks and ridges, through misty surface haze rose a dark form, a rocky black knob erupting from the ice.

It was land.

34

"I AM READY TO START!"

THE MOOD ON THE *CITTÀ DI MILANO* AT KINGS BAY WAS ANXIOUS AND agitated. The last message from the *Italia* had been in the early morning hours of May 25, reporting that they were bucking severe headwinds and would likely arrive late that afternoon or early evening. After that, communication had gone silent. Nobile's rough position at that time was estimated to be somewhere between 30 and 40 degrees north of Kings Bay, north of Moffen Island. When they did not arrive that night, or the next morning, speculation and rumors about what might have happened went around the base ship. Perhaps the aerial had iced up, thwarting communication. That was plausible, and they might still be flying around, following the wind. But one of Biagi's last messages had said they were running on only two engines to save fuel, so clearly they were concerned about fuel. Doomsayers grumbled that given the conditions, they might well have crashed into the mountains of North East Land, with rugged peaks over five thousand feet high. Maybe they had made an emergency landing on the ice, some suggested with faint hope.

When a weather report came in from the Geographical Institute at Tromsø, unease grew greater still. A strong cyclone had been ravaging the region north of Spitsbergen, blowing hard from the west and right in the *Italia*'s flight path. The gale appeared strong enough to have blown them all the way to the Siberian coast.

Umberto Nobile's brother Amadeo, in charge of base operations, stalked around the *Città di Milano* trying to get any useful information. He consulted frequently with Ettore Pedretti, the other *Italia* wireless operator who had remained behind when Nobile informed him only one operator would be needed for the North Pole flight, and it had been Biagi's turn, since Pedretti had been on the long flight to Severnaya Zemlya. Now, Amadeo and Pedretti listened for any news on Pedretti's wireless station onboard the *Città,* and frequently went to Captain Romagna to implore him to do something. Captain Romagna voiced his own concern for the plight of the *Italia* and its crew but underscored that any official orders needed to come from the defense ministry in Rome, who had told him to remain at Kings Bay for now, adding, "We can give no orders without further information on the airship's whereabouts."

The *Città di Milano,* at any rate, was unsuited for battling sea ice. Thin-hulled and with old engines, the best it would be able to do, when and if orders came, was to move farther north and east along the Spitsbergen coast, possibly close enough to intercept a message. Beyond that, airplanes—and maybe even ice-breaking ships—would be necessary.

By now, the world knew something had gone dreadfully wrong, as press reports had started to circulate around the globe. They were mostly speculative and erroneous, but the consensus was that the *Italia* had disappeared. "*Italia* Missing: Reported Down on the Ice" was *The New York Times* headline when the airship failed to reach Kings Bay, though as yet no confirmed radio message had been received from Umberto Nobile. The truth was, no one really knew what had happened.

Offers of assistance started to pour into Kings Bay from the outside world: The Norwegian government pledged air support, as did the governments of Sweden, Denmark, and Finland. The whaler *Hobby,* already in local waters and now with dog teams and sledges on board, set out to search Spitsbergen's northern coastline, though it would not be able to penetrate the pack ice. All whalers then working Arctic waters were informed of the emergency situation and were asked to be on the lookout for any sign of the airship *Italia* or its crew. Additionally, many of the most prominent Arctic explorers, including Norwegians Fridtjof Nansen and

Otto Sverdrup, and American Lincoln Ellsworth, voiced opinions about what might have happened and offered advice as to the best search methods. Nansen suggested using a British dirigible, which could remain in flight for many days.

In Oslo, Norway, on May 26, a gathering of Arctic dignitaries had come together for a luncheon to celebrate the recent daring flight by Australian George Hubert Wilkins and Norwegian-American Carl Ben Eielson. The two men had just completed the first transpolar flight by airplane, going from Point Barrow, Alaska, to Spitsbergen via Ellesmere Island, Canada. The 2,200-mile, twenty-hour flight was heralded for the unparalleled navigational skill required, and the two fliers were being fêted at a banquet in their honor.

Among those in attendance was Roald Amundsen, restless though not entirely idle in his self-imposed retirement. For some months, beginning in December of 1927, Amundsen had been carrying on a secret affair at Uranienborg with the married woman Bess Magids, the young and vibrant thirty-year-old he had met on a steamship. Amundsen was smitten by her daring and adventurous personality—she ran her own dog team in Alaska between the lucrative outpost trading stations she owned with her brothers and was an avid and deadly poker player. But she could also dress the part for society functions, donning tailor-made frocks and shoes made by the finest purveyors when such occasions called. Though their rendezvous at Uranienborg had been brief—just a few months—she had agreed to marry Amundsen. Recently, she had returned to Seattle to "settle her affairs," which presumably meant securing a divorce and getting her finances in order. She had promised to return to Norway in May or June of 1928 to wed the grizzled and venerated retired explorer.

While awaiting her return, Amundsen had agreed to attend the luncheon honoring his fellow explorers, hosted by the Norwegian newspaper *Aftenposten*. Why not? He enjoyed the company of men with a shared background of adventure and derring-do, and as the most accomplished among them, he certainly belonged. And it would be good to catch up with his companion and countryman Oscar Wisting, who shared the honor with Amundsen as being one of the only two men in the world to have reached both the South and North Poles. As toasts and good cheer and stories went round, so too did word—announced by a reporter holding a telegram—that Nobile and the *Italia* were missing. There was a hushed awkwardness when Amundsen received the news. Everyone present—indeed nearly everyone everywhere—knew of the ugly, bitter feud between Amundsen and Nobile after the heroic flight of the *Norge*. When a reporter approached Amundsen and asked for a statement, he did not hesitate. In a loud and clear voice, he proclaimed, "I am ready to start on a search for the *Italia* at once!"

35
SEPARATION

NOBILE AND MALMGREN STARED AT THE DARK, CONICAL KNOB PRO-
truding from the polar sea. It was no mirage or Arctic illusion. It must be
Charles XII Island, a tiny, mile-long landmass. If they were right—and
they felt certain they were—at least they knew their location at the moment. Malmgren thumbed through the *Arctic Pilot* and looked up with
wonder and worry on his face. Given the island's proximity on the horizon, they had drifted nearly thirty miles in the last two days, heading
southeast. Malmgren read aloud to Nobile and the others from the book:
"The principle direction of the ice-stream in North East Land is toward
the east."

Nobile asked Malmgren to predict, or at least guess, where the drift
would send them, and he responded without hesitation: "Franz Josef
Land." That was some three hundred miles to the east, and a place that
rescuers—if they were coming—would be unlikely to search for them.
Malmgren pointed to the *Arctic Pilot* book and said that Broch and Foyn
Islands—just twenty or thirty miles from North East Land's mainland,
were directly southeast of Charles XII Island, and at their present drift,
they might well be passing near them in a day or two. Malmgren proposed
that it might make sense for them to trek to one of those islands before the
prevailing drift sent them farther away from any land.

Nobile was quiet for a time, deep in thought. He asked to be brought back into the tent to warm up. Later, Malmgren and Mariano approached Nobile, once again pressing the idea of striking out for land. Their situation was dire, they argued; radio communication was not working, and they would soon drift far out into the frozen sea. They should go as soon as they came in sight of Broch or Foyn Islands.

Nobile countered with one obvious problem: neither he nor Cecioni was capable of travel. He added that he believed the radio would eventually reach someone; it was only a matter of time, and they had salvaged enough food to survive for more than a month. Also, he was not convinced the drift would necessarily drive them eastward; the winds might shift, he countered.

Malmgren conceded that the winds might shift; sadly, he knew this as well as anyone. But the prevailing currents would not. These had been long established, were even printed in the *Arctic Pilot* he held in his hands. Mariano interjected then, suggesting it might be possible to build sleds in which to haul Nobile and Cecioni; or, should that not work given their limited equipment and resources, a smaller contingent, including Malmgren for his ice experience, should go. Once they reached land, they could send for help.

Nobile said flatly that the group would remain together, and that no one was going anywhere until Biagi had made radio contact with the *Città di Milano*. Then he sighed, exasperated by these difficult decisions and his constant pain, and said that everyone should meet in the tent later to discuss the proposals. Nobile's behavior and attitude had changed since the crash. Aboard the *Italia,* he was sole commander, and he assumed the role with a firmness and certitude that everyone understood and respected. He was, after all, a general in the Royal Italian Air Force, and the *Italia* Expedition had been a military operation, one in which his orders were sacrosanct. But now he was seriously injured—literally a broken man—and although he retained command, he seemed less sure of himself, less definitive, more open to suggestion. He was at least willing to entertain options.

While Biagi had not yet been able to confirm receipt of any of his sent messages, he was managing to periodically receive radio transmissions coming out of the San Paolo (Rome) station and he transcribed them and read them out to the others. They were primarily local and regional news and sports reports, including soccer scores, of all things. But hearing something—anything—provided hope, some tangible proof that even though they were alone out on the ice, they were not alone in the world. It was encouraging to be able to hear something from civilization, but at the same time it was extremely frustrating that Biagi's round-the-clock SOS calls were going unanswered by anyone. Biagi had by now configured his remote wireless station with the transmitter outside the tent, near the aerial, and the receiver inside the tent, so that he could put on his earphones and hear more clearly with less wind noise.

Already factions and divisions were forming among the group of nine survivors. This was natural in such dire circumstances, with their fates uncertain and the thought of death ever present. Indeed, from the camp they could clearly see the makeshift mausoleum they had erected over Pomella's dead body using the remaining parts and pieces of his engine car to cover him, both to keep bears from dismembering his corpse and so they didn't have to look at him. The factions within the group occurred naturally. Nobile and Cecioni spent the most time together, both being tent-bound and unable to walk or move without assistance. Mariano and Zappi shared the same rank as naval officers, which drew them together along with Viglieri, a lieutenant commander in the navy, who gravitated toward his two superiors.

Malmgren and Behounek, both scientists, spent a good deal of time together out on the ice, contemplating leads growing around their encampment and discussing drift, currents, weather, and a safer location to move the camp. Trojani, an engineer, was rather solitary and reserved; he spent much of his time tinkering with gear and assisting Biagi as much as he could. Biagi had proved to be virtually tireless, with the emergency radio and its operation his singular focus. He was also gregarious and got on well with everyone. He had assumed with grace and dignity the weight of responsibility which fell on him: unless he could somehow make contact, they would all likely die.

In the evening meeting after their meal of pemmican glop and a bit of chocolate, Nobile asked for everyone's thoughts, including who might head toward land should they drift near enough. Mariano said that he wished to go. Zappi volunteered also, provided Malmgren could come to navigate across the ice. A smaller group, Zappi argued, would allow them to move faster. The others were all content to remain on the ice and take their chances with their commander.

Nobile considered the idea, still very uneasy about separating. He concluded that they should wait to see where the drift was taking them for a day or two, and in the meantime, they would build a sledge for a potential land attempt. Also, before anyone left, they should continue to scout for a safer camp location in the vicinity, perhaps more conspicuous and elevated, and on a larger floe.

Later on, Biagi came to Nobile with tantalizing news. He had intercepted and interpreted cryptic messages that the *Città di Milano* had left Kings Bay to search for them, and that the whaler *Hobby* was headed for Hinlopen Strait, between Spitsbergen and North East Land. If ice conditions permitted them to get there, they would be about halfway from Kings Bay to their approximate location and might well come into radio range. The news bolstered Nobile's belief that they should remain patient, and buttressed his faith that eventually they could establish communication. His main concern was that they would run the batteries down before making contact. So, for now they would wait.

"There's a bear!"

Zappi and Mariano had been out taking a sun observation when they saw the large animal just twenty yards from the tent. They hustled over, opened the tent flap door, and whispered everyone awake. "It's a large bear, just out there on the ice," Zappi whispered, with both thrill and fear in his voice. Malmgren turned to Nobile. "Give me the pistol, please.

I'm going to hunt it." After Malmgren's disturbing previous episode with the Colt, Nobile had been keeping it. He handed it over, and Malmgren checked to make sure it was loaded, then shuffled out. Nobile clasped his hand over Titina's muzzle, stifling her yips to a whimper.

Malmgren crept forward with the pistol brandished. Zappi and Mariano were just behind him, one wielding a knife and the other an axe. The bear was nosing around some gear, snorting and pawing, as the men drew closer. It appeared oblivious to them, quite possibly never having encountered a human before. Malmgren slowly raised the Colt, leveling it on the animal, then fired. The shot cracked and reverberated over the ice field. "He's hit!" Malmgren shouted. The bear trotted off, shaking its head, and the men followed. Malmgren fired twice more as he ran, and the bear dropped dead on the ice.

It was a medium-sized bear, likely a juvenile. Immediately they set to butchering the animal, and Malmgren estimated it would provide four hundred pounds of meat, and the skin could be dressed out and used for a blanket. When Malmgren opened the bear's stomach, he found little but half-digested paper with English words printed on it. The pages, they realized, had come from one of their navigation books which the famished animal must have found while rooting around the crash area. When Malmgren told the others in the tent, Nobile said they'd need to increase the daily lookout, for the thin-walled silk tent and a twelve-pound terrier provided little protection against hungry polar bears that could weigh up to 1,400 pounds.

The next morning, with everyone satiated by the previous night's ration of fresh bear meat and broth, the mood at the Red Tent had improved. Nobile thought that given their augmented food larder—the bear meat extending rations by three weeks—Malmgren, Mariano, and Zappi would be less anxious to leave. So, he was surprised when Zappi came into the tent and announced that he thought a shore party should leave as soon as possible. Nobile pressed him on the hurry, and Zappi informed him that they had drifted to within seven miles of Foyn Island. They could see it in the distance, and seagulls and other shorebirds had been spotted in the skies above.

Nobile was in a difficult position. On the one hand, it was insulting that two of his officers, Zappi and Mariano, wished to leave. That felt like a personal abandonment, and was a blow to his pride, but he tried to move beyond the emotional aspect of it and consider things practically. They believed—as did Malmgren—that with the *Città* and *Hobby* (and perhaps by now other ships) likely scouring the northern coastline, if they could get there within a couple of weeks, they would have a decent chance of being discovered there and could orchestrate the rescue of those still on the ice. But the farther their floe moved east, they contended, the farther the dangerous voyage over ice would be, so the time was now.

It was a fair argument. He understood the rationale and did see the merit in their plan. But Nobile was torn. Splitting the group worried him, and some of the others—Cecioni in particular—feared being left with only four men to take care of him and Nobile, both still essentially invalids. Nobile mulled it over. He could have issued a military order for them to stay, but he did not feel right about such an extreme measure, which could lead to intense discord at camp and possible mutiny anyway. Better to let them go with his blessing.

Nobile now oversaw the distribution of stores. Those at the Red Tent would keep all of the bear meat, since the shore party would need to travel as light and fast as possible. The food for those departing was distributed between Pomella's knapsack—found near him inside what remained of his engine car—and two cloth bundles they fashioned from torn clothes, which they would sling over their shoulders, since the attempt to build a working sledge had failed. They would carry about seventy pounds of pemmican, forty pounds of chocolate, six pounds of malted milk tablets, and two pounds of butter. For clothing, they had the lambswool flight suits they had been wearing in the *Italia,* plus some extra clothing found in the emergency bag Arduino had tossed from the airship: thick wool socks and gloves; three pairs of light leather boots each, and fur-lined hats with earflaps. There was also, to share, a large woolen blanket, compasses, one rope—useful for crevasses and open leads—and some sections of varnished canvas salvaged from the control car, on which they could sit or lie on the ice or use as windbreaks. Unfortunately, there was only one pair of snow goggles, which they would have to share. This was a problem, as snow-blindness was a very real possibility and could reduce a man to a painful, eye-searing crawl, literally. But one pair was all they had.

While Zappi and Mariano packed up the food and gear, Malmgren went inside the tent. He was in good spirits, attentive and even affectionate to Nobile, apparently relieved that he could do something active and helpful for the party as a whole. Nobile was worried that Malmgren's injured shoulder would hinder him, but the Swede said it had improved, suggesting it was only deeply bruised, and the others had offered him a slightly lighter load. Malmgren went over their proposed route with Nobile. "We shall march as fast as possible," he said confidently, "and in about two weeks, I hope to reach Cape North [North East Land]." Then, unless they were spotted and picked up by ships, they could continue on west, all the way to Kings Bay, another one hundred fifty to two hundred miles over mountains and across glaciers. Now the more Nobile thought about it, the more suicidal it sounded, but he kept his thoughts to himself.

Both men knew that even reaching the North East Land coast would be arduous. Traveling over rough, broken, and breaking ice was slow, difficult, and dangerous, and one could rarely go in a straight line for long. They would encounter open water leads they would need to cross, and high pressure ridges. But Nobile just nodded and smiled. "And then what?" he asked.

"Then I will come back and look for you myself, with . . . airplanes . . ."

The two men were quiet for a time, both comprehending the gravity and implications of the separation. Malmgren had been with Nobile on the historic *Norge* flight, on the recent record setting flight to Severnaya Zemlya, and had survived together their recent dirigible crash in the *Italia*. It was a lot to take in, the understanding that they might well never see one another again. Malmgren broke the silence, asking Nobile what message he had for Italy.

"Tell them that my comrades and I are staying here calmly," Nobile answered. "If we can be saved, so much the better; if not, we will wait serenely for death, satisfied that we have done our duty. Only see that they look after our families."

Malmgren listened to the general's wishes and nodded gravely.

"All right, General," Malmgren promised, adding, "In any case, re-member that the greater number of lost expeditions have been saved at the last moment."

Nobile managed a weak smile.

Those remaining wished to write letters to their families for the shore party to take, and did so while Malmgren, Mariano, and Zappi packed up all of their gear. Nobile scribbled hurriedly, dashing off seven full pages to his wife, Carlotta, including advice on young Maria's education. In closing, he tried to give her solace in the event he did not return. "Perhaps God wills that we shall embrace each other again one day; that will be like a miracle. If not, do not mourn my death, but be proud of it. Be certain that I have done my duty quietly to the last." Nobile took out the picture of Maria found among the scattered wreckage. He gazed into her plain-tive eyes, then continued writing, now directly to her: "You must keep Mummy from crying, if I don't come back again. Titina is perfectly happy here, but perhaps she would still rather be home."

By the time Nobile finished, all the others had written to family mem-bers as well and bundled the letters together. Everyone gathered in the tent to bid one another farewell. Mariano and Zappi came forward and embraced Nobile, each in turn. Everyone present had tears in their eyes. "We will march swiftly and bring help," Mariano assured Nobile and the others.

Outside the tent, the three departing men hitched their bundles onto their backs while others helped secure the straps. Then Malmgren,

Mariano, and Zappi started off in single file, tromping across the crusty ice in the direction of Foyn Island. The remaining survivors watched them go, the three forms growing smaller and less distinct in the distance until they were just dark specks, indistinguishable among the hummocks and boulders as they melded into the horizon.

36
CONTACT

ON THE NIGHT OF JUNE 3, 1928, NIKOLAY SCHMIDT, A RUSSIAN farmer, was sitting by his wireless set at his home in the small village of Archangelsk, on the White Sea. After working his fields all day, he liked to spend his evenings on his radio, communicating with the outside world. On this night he heard something that caught his attention. It was a message, cryptic and choppy, but he jotted down what he could decipher: "Ital . . . Nobile . . . SOS . . . SOS . . . SOS . . ." He had been keeping up with the news and knew about the lost *Italia.* The communication seemed legitimate despite being partial and containing no location. Schmidt immediately contacted the Soviet Embassy, who forwarded it on to authorities in Rome. At least some members of the *Italia,* it appeared, were still alive.

The miraculous news made its way around the world, and rescue efforts—which until now had been mostly in planning phases—ramped into high gear. Officials and explorers met more than once at the Norwegian Defense Ministry, in central Oslo, finalizing their tactics, securing aircraft, and communicating with Rome and the *Città di Milano,* which had by now steamed from Kings Bay to near Danes Bay. Because the position of the *Italia* survivors remained unknown, a large swath of the Arctic would need to be covered, including northern Svalbard, Franz Josef Land, Severnaya Zemlya, and all along the northern Siberian coastline.

"The area to be searched is enormously vast," Fridtjof Nansen pointed out, adding with typical understatement, "It won't be an easy undertaking."

Norway's Hjalmar Riiser-Larsen, who had been instrumental on the *Norge* flight, was particularly motivated, having come to know Nobile, Malmgren, and many of the crew intimately. He had shared both travails and triumph with them. He volunteered to go immediately and would travel by airplane or airship—it didn't matter to him. "Give me a machine such as the situation demands," he said to a reporter, betraying a bit of bravado, "and I will straight away begin a flight to find the lost men. With me I will take two guns and plenty of ammunition . . . If I am compelled to land, the rest can be left to me. The bears and seals will supply me with abundant food and a protection against the cold; and next year you can come fetch me."

Even Walter Wellman, who had pioneered Arctic airship flight in the same region two decades earlier, weighed in with an opinion. In an interview with *The New York Times*, Wellman said, "Nobile's provisions must be running low by now. About the only chance they have of rescue is by airplane, and even a plane would encounter great risk in attempting to land on the pack. But it might at least drop supplies and food."

A multipronged international effort to locate and rescue the survivors was at last underway: The Italians committed two flying boats and a small contingent of Alpine Ski Troops to search from land; the German government pledged a pair of Dornier-Wal flying boats; Sweden dispatched experienced air force pilots in special float planes; the French provided airplanes and skilled pilots; and the Soviet Union sent three icebreakers—the *Georgiy Sedov*, the *Malygin*, and the *Krassin*, all loaded with ski planes on board. Of these formidable ice ships, the *Krassin* was known to be the most powerful icebreaker in the world and had previously performed an Arctic rescue when it saved eighty-seven men aboard a ship that was broken down and trapped in pack ice. All counted, sixteen ships and twenty-three airplanes from eight countries—involving approximately 1,400 men—were committed to the largest single search and rescue mission ever attempted in the Arctic.

The international outreach was impressive, though the effort was not exactly coordinated nor based out of a central command. There was cooperation among the nations involved, to be sure, but in respects each nation was also acting independently, vying for the glory of the rescue and the prestige and national pride that would come with it. The Italians should really have been spearheading the search since it was after all their lost airship and their expedition but compared to the other national governments in the hunt, and especially compared to the Scandinavians, the Italians had the least Arctic experience of anyone.

Roald Amundsen was also preparing to head north. In his mind, given his credentials and experience, it was obvious that he should be the leader of the Norwegian rescue effort. But he was not chosen as commander of the Norwegian naval expedition—that honor went to Riiser-Larsen, who was appointed after Mussolini had personally sent a telegram to the Norwegian government expressly asking that Amundsen not be officially involved in their search. Mussolini still harbored ill-will toward Amundsen for his constant public disparagements of Nobile—and by association, Italy—after the *Norge* flight. So, the disgruntled Amundsen needed to find another way to join the search, and he reached out to some of his considerable network of international contacts. A prominent Norwegian businessman living in Paris eventually managed to secure a French naval flying boat and a crew of four for Amundsen. The seaplane was in France, and it was agreed that when Amundsen was ready, the French pilot and crew would fly to Norway, collect Amundsen, and continue north.

As Amundsen was preparing and finalizing details, he was visited at his home by Italian journalist Davide Giudici, who wanted to know why, after all the bad blood with Nobile, Amundsen would risk his own life on an Arctic search and rescue mission. Amundsen sat at his desk, and he paused, considering the question before answering carefully. "There is a sentiment of solidarity which must bind men," he said, "especially those who risk their lives in the cause of science." He turned and looked across the room, his eyes taking in the pictures on the wall; one of the *Norge* in flight, and another a large map of the polar regions. Then he went on. "Before this sentiment our personal resentments must disappear. Anything that has disturbed my relations with General Nobile has been forgotten for some time. Today I see one thing only: General Nobile and his companions are in danger, and it is necessary to do everything that is humanly possible to save them."

Amundsen's tone was earnest and genuine. There was something that bound fellow explorers together, a camaraderie whose only equivalent might be soldiers who had shared foxholes and the heat of enemy fire together. From the moment he had heard about Nobile's disappearance, Amundsen had been singularly focused on joining the search. Critics could, and would, say what they wanted. It was true that finding Nobile himself would be a fitting, even ironic, end to their shared story, and he could not argue that there would be glory and heroism in it, an exclamation point at the end of his storied Arctic career: the polar legend rises from retirement and swoops in like an Arctic knight to rescue his archnemesis from the frozen wastelands. One really could not script a better ending.

Amundsen went on to say, "In the next few days I shall make a choice of the personnel, mechanics . . . and pilots who are to accompany us. I shall have under my orders the very best men, with whom, I am certain, we shall succeed. If other expeditions arrive, we will willingly collaborate with them. The question is not who will arrive first or who will do the most . . ."

With those sentiments out in the clear and on the record, Amundsen rose and walked to the hallway to listen to his wireless set transmitting the latest news. There was nothing new being reported, so he made his way back toward his desk. The journalist noticed a small model airplane hanging from a beam, and he asked Amundsen about it. The model was a replica of the Dornier-Wals N25, in which he had flown with Riiser-Larsen attempting to be first by air to the North Pole. Amundsen regarded the airplane, becoming lost in memory of that near-death experience on the ice, of hacking out a runway for weeks, and finally flying back to salvation.

Finally, he looked away from the model airplane and cast his gaze out the window to the fjord, and beyond, to the north. "Ah, if you only knew how splendid it is up there," he said. "That's where I want to die; and I wish only that death will come to me chivalrously, will overtake me in the fulfillment of a high mission, quickly, without suffering."

37
STAGGERING

THE THREE MEN STAGGERED WESTWARD OVER THE SHARP AND broken sea ice, tripping and stumbling along in ragged single file. Malmgren remained out front, with Zappi some yards behind him and Mariano at the rear. Sometimes they trod over rock-hard humps of ice buried by snowdrift; other times, they came to leads of water—like creeks or rivers—too wide to cross, and they had to follow these, one direction or another, sometimes for a mile or more, to find a place narrow enough to cross. Their feet were soaked, their eyes burning from glare. Malmgren's bad arm hung uselessly from his side, and he lurched along in silent misery, often stopping to kneel and gulp for breath and squint out across the sastrugi—the wind-blown ice ridges stretching like dunes as far as he could see.

The land in the distance, which they believed was Foyn Island, seemed to be getting farther away instead of closer, though they had been plodding along for days. It was hard to know what day it was, or how long since they had left the Red Tent, because of the constant sunlight, with no boundaries between day and night. All three men were wet, cold, dehydrated, and exhausted, but Malmgren was in the worst shape. One shoulder hung lower than the other, and as he reeled ahead, he was bent over, hunchbacked, hissing tight-lipped with each painful step.

Malmgren's sufferings were physical, mental, and emotional. His shoulder popped in and out of the socket with every stumble or leap across a lead, and it had become so useless that he had been forced to allow the others to carry nearly all of his share of the provisions. Inside he felt damaged, too, with bolts of pain shooting from his abdomen around his hip to his lower back. He must have had internal organ injuries from the crash—maybe a kidney or his spleen—and he had started checking his urine for blood. He was deteriorating mentally and emotionally too. As they trudged for twelve to fourteen hours a day (whatever a day was), his mind had time to relive his faulty wind predictions, and he could not keep from blaming himself for their predicament. If he had only agreed with Nobile and not turned back into the wind, likely they would not have crashed. Now, all he could think about were his failures. He had insisted he lead the shore party, and that they could make an average of eight to ten miles a day. But the others were forced to carry his load and he was slowing them down to just a few miles a day. Stumbling along, his thoughts at times crystalized and clarified, and he envisioned his home in Uppsala, and his lovely fiancée, whom he had promised to wed when he returned. In his lowest moments he doubted he would ever see her again.

After a few days of constant struggle, the weather worsened, with blowing wind obscuring the outline of land to the west, then closing in on them with such fury that it was difficult even to read the compass at arm's length. Water leads appeared like black oil slicks coursing through the pack, and they used their rope to tether themselves together some yards apart, sloshing through the slush to the other side. In the diminishing visibility Malmgren tripped over a snow-buried ridge and fell onto his side with a gasp. He rose slowly, plodded ahead, and fell over again, this time lying in agony for a time before rising to his knees, then standing, hunched over at the waist.

Mariano and Zappi came to him, helping him stand upright. Icicles hung from all of their mustaches and beards in what had become a violent blizzard. Wind-driven graupel—tiny snow pellets—lashed and stung their faces. Zappi located a nearby ice ridge and climbed over to its protected leeward side, then took out his axe and began hacking away at the solid ice wall while Mariano helped Malmgren over there. As Zappi chopped off ice chunks and dug a depression, Mariano and Malmgren banked up a wall on one side and used the varnished fabric they had brought with them as a floor. After a while Zappi had carved a hollow that, while not a full-fledged snow cave, was deep enough for them to crawl into for protection against the driving wind and snow. They covered their legs with another sheet of the varnished canvas.

It was too windy to light the stove, so they sat shoulder to shoulder and gnawed on chunks of pemmican, straining to swallow since neither could they melt ice for water. In this way, they shivered and writhed for the next day and a half, waiting out the storm, only rising occasionally to relieve themselves or try to get some bearings.

At last, the winds relented, and the visibility improved. Rising from their ice pit, they shook snow from their hoods and coats, and stamped around, trying to get warm. Zappi and Mariano rolled up the varnished cloth and put away the axe and bundles of food, and they all scanned the horizon. A thin gray line seemed to hover indistinctly out in the distance, and they all cursed as each in turn checked the compass; their drift had taken them some miles southeast—it was impossible to know how many—away from Foyn Island. All they could do now was set a course toward the line in the sky and slog onward.

For what must have been two days they lumbered in a zigzag course bearing generally west, uncertain of any progress they might be making, time marked only by the cadence of their breathing, stops to rest and eat pemmican and chocolate and drink meager thermos capfuls of meltwater. Thirst crazed and weak, Malmgren began wobbling and weaving, careening like a drunk until he tumbled and fell and lay splayed out on his back, muttering incoherently.

Mariano and Zappi knelt beside him. They gave him the last of their water and tried to sit him up, but he resisted. "You two go on. Leave me here," he husked.

"Finn," said Zappi brusquely, "Get up off the ice. This is no way to act."

Mariano slung off his knapsack and filled their empty fuel can with snow, then managed to light the small oil stove and boil a thick sludge of melted snow and pemmican. They poured some in the thermos lid and held it to Malmgren's quivering lips. He sipped at first, then gulped it down, and they gave him more and drank a potful themselves. After a time, Malmgren's eyes blinked and brightened, and they helped him into a sitting position. At last, he was revived even more, and he apologized to them and said, in a voice dry and reedy, that he could go on.

With effort, by one man holding his good arm and the other hoisting him by the waist, they got him standing. Malmgren looked at the compass, pulled the snow goggles down over his eyes, and started limping and staggering once more across the labyrinthine pack ice.

38

"THEY'VE HEARD US!"

TITINA WAS YAPPING AND SNARLING OUTSIDE THE RED TENT. NOBile sometimes let her roam around outside freely, especially at mealtimes, so she wouldn't get in the way, knock over the cookstove, or annoy them with her begging. Sometimes she barked at geese winging overhead, or seagulls, but this sounded different, urgent and protective. As she yipped and growled, Trojani, who had been at sentry post, stormed into the tent. "Give me the pistol—there's a bear." Nobile grabbed the pistol from the center tent pole, where it had been hanging since Malmgren had left, and handed it to Trojani, who bolted back outside.

Everyone but Nobile and Cecioni scrambled out of the tent after Trojani. They watched in amazement as a giant polar bear lumbered away, kicking up snow, with little Titina snapping at its heels. Trojani was in pursuit, firing as he ran, but he missed, and the bear dove into a big lead of water and swam off, disappearing into the ice fields on the other side of the lead. The pursuers returned to the tent to finish their measly meal, and Nobile patted Titina's brown face as she wagged her tail. He was relieved she had not been killed with one swat of a massive paw. At first everyone laughed at the absurd image of Titina chasing after a massive polar bear, but then grew bitter when it occurred to them that she had just chased away perhaps a month's worth of food.

The mood at the Red Tent had been mixed after the departure of Malmgren, Mariano, and Zappi. There existed a heightened level of hope and expectation that the men would make it to land and send help, but that optimism was tempered by the reality that it would likely be weeks, at least, before they struck mainland North East Land. After that, it was anyone's guess whether they would be located, or by some miracle trek all the way to Kings Bay. Nobile calculated that the shore-bound group would run out of food in three weeks.

Nobile wanted those who remained—himself, Cecioni, Biagi, Behounek, Viglieri, and Trojani—to continue on with some level of routine. Biagi maintained his round-the-clock radio vigil, still with no luck. Viglieri took charge of the inventory of stores, and even devised a clever method of roping together all the containers of pemmican, chocolate, and malted milk tablets so that, in the event that ice breakup occurred beneath or near the tent, they could grab the rope and keep the vital food from plunging into the water. Viglieri also kept recording solar observations with a sextant to determine their position as best he could.

Behounek and Trojani shared cooking duties, which, with Malmgren gone, included the never-ending task of finding old ice for melting. They worked well together and tried to vary the cooking of the bear meat they had left from the animal Malmgren had shot, alternating between boiling and frying, though their butter was almost gone. Sometimes they mixed some bear meat in with pemmican and made a "bearican" stew.

Cecioni had initially been the most reluctant to see the party split. He had grave concerns about seeing three of their ablest members leave and feared that would ensure his death on the ice. But after some days his spirits improved, and though he could not walk, he was able to pull himself around by his arms, dragging his broken, splinted leg behind him. Trained as a dirigible technician, he was very handy. Among the found items were some sewing materials: needles and fabric and strips of leather. Cecioni started sewing over-slippers to sustain the life of the men's boots, which were wearing out from walking on the hard, sharp ice. Viglieri had very large feet, and Cecioni fashioned a pair of over-shoes especially for him using the tough, waterproof canvas of a tool bag that had been found.

One thing that improved the mood in camp was that they had drifted closer in proximity to Broch and Foyn Islands. Daily sightings of land raised their spirits, and one day they estimated that Foyn—the nearer of the two—was within five miles. They began referring to it simply as "The Island." They scanned for hours, sharing the couple of pairs of binoculars they had, also glassing for any sign of the shore party, but all they saw was the rough outline of rocks and snow. They were so close to land that seagulls started coming to camp in large numbers, lighting on water leads and on hummocks nearby, filling the air with a cacophony of cries. "These signs of life," Nobile wrote in his notebook, "cheered us up immensely. We felt less alone. We thought that if there was life around us there was also a chance of prolonging our own, when one day our provisions were exhausted."

Some seals had also started appearing, sunning themselves on the ice beyond a lead not far from the tent. But every time one of the men grabbed the Colt and stalked them, the seals slid from the ice, disappearing below the water's surface. One day while Biagi was out working on the receiving apparatus, he saw schools of fish swimming up the lead closest to the tent. Cecioni bent some of the larger sewing needles into hooks and tied them onto stout thread for line, but no one managed to catch any of the fish, which someone thought were called tomcod.

A consolation of having the shore party gone was more room inside the tent. It was still cramped, but in the evenings after meals they would tell stories, sing Italian folk songs, and Nobile would try to bolster morale by reminding everyone of that before the crash, they had accomplished the most significant scientific aerial expedition yet, fulfilling the mission bestowed upon them by God and country. As he said these words, his eyes fell on the tent pole, near the top of which they had tacked the picture of *Madonna of Loreto*. The image, by Italian master Caravaggio, depicted the barefoot Virgin holding her naked child in a doorway between two kneeling peasants, their palms pressed together in supplication. Nobile often looked to the picture of the *Madonna of Loreto* as he prayed. Beyond its spiritual use, the tent pole served as a place to hang the Colt for easy access, and someone had also fashioned pegs along the pole to hang boots and clothing to dry. Mid-pole, they had also affixed a small calendar to mark each day on the ice.

By the morning of June 6, after twelve days on the floe, Nobile was growing deeply concerned. With good visibility for an extended period, they had noticed "The Island" receding behind them. They had drifted past it, and now as they scanned to the south, they caught glimpses of Cape Leigh-Smith, the northeastern-most point of Svalbard's mainland. Their movement confirmed what the *Arctic Pilot* book said—the principal direction of the ice stream was toward the east. Nobile wrote in his journal: "Where should we end up? Perhaps in a few days we should reach Cape Leigh-Smith, pass it, and drift inexorably eastwards away from land. All possibility of help reaching us in time—even by chance—would vanish."

Nobile had other worries, too. Large rifts and fissures were forming in the ice all around the tent, threatening to dislodge them and leave them on a tiny, crumbling island that could soon be subsumed into the sea. And there was the issue of the radio batteries, which he and Biagi both knew could not hold out indefinitely. Given these realities, Nobile made some difficult decisions. He wrote a message in Italian, French, and English: "S.O.S. *Italia,* Nobile. On the ice near Foyn Island, north-east Spitsbergen . . . Impossible to move, lacking sledges and having two men injured. Dirigible lost in another locality." He handed it to Biagi: "From tomorrow onwards," he said, "you will send it out every day for an hour on end."

His next decision was even more drastic. If their drift took them quite near Cape Leigh-Smith, he would send the four able-bodied men over the ice toward land. For them, he believed, it might be their last, best chance to survive. He and Cecioni, unable to walk, would remain at the Red Tent and await their fate: either rescue or death.

That evening, with everyone huddled around after their meal, Biagi sat cross-legged, his headphones cupped over his ears beneath the flaps of his fur-lined hat. He was writing dispassionately, transcribing yet another mundane news report from San Paolo, when suddenly his countenance changed. His eyes widened and he looked to the others with amazement, now writing furiously. "They've heard us!" he shouted. "They've heard us!" He started scribbling fast in his notebook, then handed it to Nobile: "The Soviet Embassy . . . has informed the Italian government that . . . an SOS from the *Italia* . . . has been picked up by a young Soviet farmer, Nicholas Schmidt, at Archangelsk . . . night of June 3 . . ."

Nobile read the message aloud and a cheer went around inside the tent. Then each man read the message to themselves, as if needing proof that it was true. Nobile doled out celebratory tablets of malted milk—three each per man—and a dram apiece of pure alcohol as libation, which he had been saving to use as medicine. The implications of the message were profound. It had been three days since the Italian government had learned that at least some of them were still alive, and in dire need of help. Surely by now action had been taken.

Despite what he knew would be a drain on the radio batteries, Nobile instructed Biagi to begin sending messages even more frequently, including their updated position each time. Their location proved difficult to pinpoint, as another storm had blown in, but Viglieri did his best to get some sun shots and at least they thought they had reasonably accurate coordinates to transmit.

On the morning of June 8, Biagi was alternately transmitting and listening intently, almost as if he was trying to will a real-time two-way communication into being. He had altered the message more than once, settling for now to report their location as "about 20 miles from the northeast coast of Spitsbergen." The last accurate solar observation, and their sighting of Cape Leigh-Smith, put them somewhere in that vicinity.

That evening, everyone huddled around Biagi, pressing in close to the radio receiver listening to the San Paolo newscast with renewed interest. When the standard press reports concluded, Biagi started writing furiously, pressing his other hand tight against his earpiece. Suddenly he grinned broadly and shot a look at Nobile.

"They are calling us!" he exclaimed. "They are calling us!"

Everyone was silent, waiting for Biagi to say something else. Their breath billowed out in plumes like smoke in the freezing air. "The *Città di Milano* heard you well this morning," Biagi read out loud, "and has received your coordinates." Nobile was so thrilled that he rewarded Biagi with an entire chocolate bar, and the rest of the men applauded Biagi for his tireless efforts on the radio.

All the next day the Red Tent had frequent two-way communication with the *Città di Milano,* and late on the night of June 9 they received the most uplifting message they could imagine: "Be ready to make a smoke signal. Aeroplanes will be . . ."

It was only a partial message, but it was enough. Airplanes were coming for them.

39

"LEAVE ME."

THE CONTINUOUS DAYLIGHT AND THE BEHAVIOR OF THE LIGHT ON the ice confused and confounded Finn Malmgren. Sunshine occasionally pierced the cloud cover overhead in concentrated beams, bolts of orange-ochre light glazing the pack ahead and illuminating the mottled surface enough for him to weave through the broken rubble in a relatively straight line for an hour or two. When the clouds knitted together, the light diffused into a dull oyster-gray, a monochromatic sameness enveloping everything: the ice underfoot and sky above seemed compressed into one continuous plane. He gasped and coughed as he plodded onward, occasionally glancing at his compass through the dark lenses of his snow goggles, then adjusted his course, tripping over pressure ridge debris as he leaned headlong toward an imaginary island, for he hadn't seen land in days.

Malmgren had somehow managed to revive and lead the trek for the last two or three days. Who knew how long it had been? In the incessant light of the midnight sun they had lost track of time and could not discern night from day. Time became measured by the number of steps between stops, the number of steps to get across leads, the number of leaps through knee-deep saline slush. But now, Malmgren sat on a hummock of ice, its top wind-shorn smooth as a bench, and watched as Mariano and Zappi moved past him and took over the lead. He bent forward, head in his gloved hands. His feet were numb, like dead stumps. He scooped some fresh snow from his ice bench, put it in his mouth, and closed his eyes. As it melted, he swallowed slowly and imagined it was a long draught of water from a tall glass.

When he rose, he could no longer feel his feet, and though he tried to catch up with the others, he faltered, wheeled and spun, then tumbled onto his back and lay there, staring up at the endless sky. He heard voices, footsteps crunching near, then dimly recognized Zappi and Mariano hovering above him. "I can't go on," Malmgren husked, his lips cracked and bleeding. "There is nothing left for me but to die."

The two men knelt beside Malmgren and tried to lift him to his feet, but his body slithered from their grasp. He lay there for a time, his breathing a low, tortured wheeze. When they implored him to get up, he pushed himself into a sitting position, the pain in his shoulder causing him to yelp. The others offered him a chunk of pemmican, but he waved it off and reached down and untied one of his boots. With difficulty, he pulled out his foot and took off his soaked woolen sock. His foot was purplish black, foul and necrotic with frostbite. They tried to rub circulation back into it, but he winced away, wriggling the sock back on. Both feet were like that, he told them.

"Leave me," he said firmly, "Let me die peacefully . . . But you must continue at all costs. You must save the others."

Mariano and Zappi looked at each other. They had nothing with which to make a litter, no sledge. They could not possibly carry Malmgren; they were weak and famished themselves. For a while no one spoke. Ice fractured somewhere in the distance with a grating sound as one floe ground into another. Malmgren held out his compass. "Give this to my mother," he husked, "as a remembrance. And tell them in Sweden why I did not return." Then he took off his snow goggles and handed them to the others and closed his eyes.

Mariano and Zappi spread out a length of varnished cloth in a depression next to a low ice ridge and rolled Malmgren onto it, then placed a few rations of pemmican and chocolate beside him. Zappi knelt by Malmgren, took off his gloves, and pressed his index finger into the snow. He made the sign of the cross over Finn Malmgren's forehead, then stood.

The two men regarded their companion for a time, then turned away from him and started footslogging toward the southwest. They tramped along in silence, each man pondering the weight of their decision: they were leaving their comrade Malmgren to die. When they stopped to rest, they looked behind them. They could still see the low ridge where they had left the Swede, but he was obscured from view. They ate some pemmican and drank water, waiting to see if by some miracle Malmgren would rise and follow them, but he did not. So they kept waiting, and when the light was such that they believed it was evening, they rolled out their ground cloths and went to sleep.

When they awoke, they had no idea whether it was day or night. They believed it was June 15 or 16 and estimated that they had been in the same place, less than a mile from Malmgren, for some twenty hours. Zappi stared back across the pocked ice plain, trying to will Malmgren to rise and follow them. But there was nothing but the sound of rupturing ice and a low-slung wind sweeping snow across the seascape.

They rose, packed up their dwindling stores, shouldered their rucksacks and trudged slowly away, finally leaving Finn Malmgren for dead.

This time, neither man looked back.

40

THE LAST FLIGHT
OF THE LAST VIKING

ROALD AMUNDSEN'S HEART WAS HEAVY AS HE BOARDED THE NIGHT train in Oslo, bound for Bergen. He forced a smile and waved to the large crowd that had gathered at the platform to send off the "White Eagle of Norway" as he chivalrously joined the hunt for Umberto Nobile and the lost airship *Italia*. As the train pulled out of the station, his eyes were wet with tears.

It had been an emotional ten days. On June 6, as he finalized preparations for his rescue flight, he had agreed to a visit from his dear friend Sverre Hassel. Hassel had been with Amundsen and three other Norwegians when their small team became the first men in history to reach the South Pole in 1911. Hassel had been ill recently—having just spent a month in the hospital receiving treatment for a heart condition. Amundsen had looked forward to spending a few hours with his friend and hoped that the fresh country air of Uranienborg would do him some good. But on the afternoon that he arrived, while strolling in the gardens just outside the house, Sverre Hassel suffered a massive heart attack, collapsed, and died before they ever had a chance to speak. Amundsen had been inside and heard the commotion and had hurried downstairs to find his comrade dead. Less than a week later, Hassel was buried in Oslo, and Roald Amundsen had been there to lay a wreath on his grave.

Sverre Hassel had been just fifty-two, four years younger than Amundsen. His sudden death in Amundsen's front garden was a very real reminder to Amundsen of his own mortality. Amundsen had had some recent health scares of his own and had received treatment in London for "heart trouble." He had also had a cancerous tumor removed from his thigh, followed by radium treatment. So, there was plenty of heaviness weighing on him as the train chugged through the valleys and mountains toward Bergen, where he would meet the French pilot René Guilbaud, board the prototype Latham 47 "Flying Boat"—only the second ever built—and fly north to Svalbard.

There was also a good deal to look forward to. He was, for the first time in years, financially solvent. Knowing that the rescue mission was dangerous, Amundsen had recently sold all of his medals to a prominent Norwegian businessman who had agreed to bestow the medals to the nation. It rendered Amundsen entirely free from debt, something that was deeply important to him. Also, he had been corresponding frequently by telegram with Bess Magids in Seattle, and she was preparing to board a steamer for Norway. Amundsen had arranged for her to be received and looked after "when she in his absence arrived," and they would be married upon his return from the Arctic. He had told her—and his family and the press—that he expected to be gone about two weeks, perhaps a month at the most. Umberto Nobile's position on the ice was now approximately known, and Amundsen intended to find him and the other survivors and deliver them from the ice. Duty—and perhaps polar immortality—were drawing him north one last time. It would be his final act.

In Bergen, Amundsen convened with the pilot of the Latham 47, Captain René Guilbaud, and the rest of the French crew: Albert Cavelier de Cuverville, second-in-command; Gilbert Brazy, mechanic; and Emile Valette, wireless operator. Amundsen had again enlisted, and was traveling with, Leif Dietrichson, his trusted friend and the pilot of the N24 during the 1925 attempt on the North Pole. After going over their flight plan and gear, the six men boarded the Latham and flew the 750 miles to Tromsø, arriving there on the morning of June 18.

The port of Tromsø was busy with ships preparing to head north, as well as seaplanes and their rescue crews from Finland and Sweden. Amundsen spoke with these pilots about their plans, comparing the known information: Nobile and five other survivors were encamped at a small tent which they had dyed red. Three others from Nobile's party had left him more than two weeks before and headed on foot toward Broch and Foyn Islands, hoping thereafter to make it to the mainland coast. But Amundsen knew that on the drifting pack they might be anywhere. Nobile's messages reported that there had been six men still in the *Italia*'s envelope at the time of the accident; they had last been seen in the air, floating toward the east.

Amundsen knew that locating any of these parties would be difficult enough, but successfully landing on the pack, or in leads nearby, and then taking off again with survivors presented the real challenge. As he carefully assessed the Latham 47, he pondered the biplane's strengths and limitations. Range would not be an issue; it was designed for long-range flights over the Atlantic. Fully fueled, it could fly a distance of 2,500 miles. It was powerful enough, with twin 500-horsepower engines mounted back to back driving two four-bladed propellers, giving it a top speed of 80 miles per hour. It had a ceiling of 13,000 feet, so flying directly over the high mountains of Svalbard would not be a problem. It had been designed to land on water, with a sturdy waterproof fuselage and floats mounted beneath the tips of each bottom wing, so it could conceivably also land on snow and ice. Weight and cargo were factors of concern, however. The Latham 47 would start out fully fueled and be carrying a crew of six—so it would be heavy and would struggle lifting from the water. Once Amundsen found Nobile, it would probably be necessary to leave members of his own crew on the ice and shuttle Nobile and the survivors one or two at a time back to safety at Kings Bay, or to Danes Island. He would figure that out once he located Nobile.

As they prepared to depart, one of the Swedish pilots suggested that they all fly north together, in a convoy. It would provide some safety since they could communicate with one another. The route from Tromsø to

Kings Bay was six hundred miles across the Barents Sea, with tiny Bear Island—just sixty-nine square miles—the only land in between for the estimated seven-hour flight. No airplane had yet successfully flown this stretch of sea, so there was plenty of danger and uncertainty involved. But Amundsen dismissed the idea of a convoy, telling the Swedish pilot that he preferred to go it alone. They said that was fine; they wanted to wait for more favorable winds anyway.

At 4:00 P.M. on June 18, 1928, Amundsen and the crew of five climbed into the Latham 47. Amundsen's friend of two decades, Tromsø local Fritz Zapffe, was there to watch him go. Zapffe had served on support crews during a number of Amundsen's missions, and he regarded Amundsen intently as the great explorer buckled his harness and stared resolutely out across the mouth of the fjord, flanked by brooding mountains. "I shall not forget the expression on his face," Zapffe wrote of that moment, "sitting astern, something extraordinary and resigned was over him. It appeared that nothing concerned him . . . and yet it was maybe all about him. He sat quietly just looking at me."

The gray-blue biplane taxied out into the fjord. The skies were clear and bright, the water glassy and reflective as a mirror. Heavily laden, the flying boat whined and sputtered atop the water for a great distance, spitting sea spray behind as it struggled to lift off. Finally, it ascended, flying low above the water for a long while before it rose and banked northwest.

Sometime later, a fisherman working about forty miles off the coast saw the Latham 47 and watched as it approached a fog bank hovering on the horizon. He reported that "the machine began to climb, presumably to fly over it, but then it seemed to me she began to move unevenly but then . . . she ran into the fog and disappeared before our eyes."

At 6:00 P.M., the Geophysical Institute at Tromsø received a message from the wireless operator on the Latham 47: "Nothing to report, all's well." An hour later, another message arrived, this one foreboding:

DO NOT STOP LISTENING. MESSAGE FORTHCOMING.

No message ever came.

41

DELIVERANCE

jani, running from his sentry post, called out in alarm. High winds churned up floes all around them, and Trojani told Nobile that a lead thirty feet wide had opened just yards away, and ice sheets were dislodging everywhere. Their own floe looked like it might soon be pulverized in the roil, crushing or drowning them all. They must move camp immediately.

Trojani had scouted around and found a safer location. It was only forty or fifty yards away, at the spot where Malmgren had shot the bear. The ice there seemed stable and connected to a much larger and less fragile floe. Nobile and Cecioni crawled through the tent door, dragging their broken legs behind them, while the four able men began moving stores and equipment to the new camp. Snow fell in flurries, and Nobile regarded the scene: "The ice on which we had been living for eighteen days was churned up and dirty. Here and there were puddles of water, and everywhere there was wreckage: pieces of twisted tubing, rags, broken instruments. A short distance away some reddish streaks on the snow revived the memory of the catastrophe." The long red lines left by the dye were a stark reminder of the trauma they had all endured and continued to endure.

Nobile tried to crawl his way over to the new camp, but when he reached a deep crevasse he stopped, and the others laid him on a make-shift litter and carried him over to the place they were now calling "Bear Floe," since the polar bear's carcass was nearby, lying there poking out of the snow. It took a few hours, but eventually they had moved Cecioni, too, plus all the provisions, the wireless, and set up the tent. The move had been excruciating for Nobile and Cecioni, whose fractured limbs were agitated as they were transported across the broken ice, but they bore it without complaint. Once back inside the Red Tent in the new location, Nobile wrote in his journal: "All things considered, the pack was less terrible here than in the spot we had left, chiefly because there were no traces of the catastrophe." At the first camp they lived among heaps of wreckage, a daily reminder of their horrible crash. Now, they were far enough away from the scene that it was indistinguishable. It was also good to be out of direct sight of engine mechanic Pomella's remains, which had recently been scavenged by a polar bear.

Biagi got the wireless operating on "Bear Floe," with the transmitter outside the tent and the receiver inside, and they listened carefully to the daily news reports. From these they learned that a number of airplanes—including one piloted by Riiser-Larsen and another by an Italian pilot named Umberto Maddalena—would soon be in the vicinity and flying grids over their reported location. Nobile ordered round-the-clock sentry watch, and had the men prepare to light signal fires with rags and canvas soaked with petrol and bear grease. They had also found a few flare guns among the detritus left in the *Italia*'s wake, and Nobile told everyone to have these ready as well.

After the ice turmoil that had forced their move, the weather calmed, the clouds parted, and the sun shone bright and warm for days on end. The glorious weather improved the mood at "Bear Floe," and Nobile was encouraged because it also meant perfect flying conditions. He was right. One afternoon, while Behounek was on sentry duty, Nobile heard him hollering, "Planes! Here are planes!" Everyone shuffled out of the tent and looked to the sky. About two miles distant, they could see two small black dots cutting across the swath of deep blue. Cecioni, poking his head out of the tent, could not contain his elation. "We're saved!" he cried out.

Trojani rushed to light the signal fires, and black smoke billowed forth, but in the calm air it hovered along the surface ice in thin wisps instead of rising up in thick plumes, as they had hoped. The airplanes kept their course, moving away from them, and Viglieri fired two flares into the air. But the two airplanes banked away westward and out of sight.

Everyone's spirits sagged. They had been so very close. Nobile immediately dictated a message for Biagi to send to the *Città di Milano*: "Today we saw two aeroplanes coming in our direction, a mile or two south of us, without reaching us . . . come back this evening before the weather breaks up. Instruct pilots to follow same course but continue along it for two miles northward. We will have smoke signals ready."

But the airplanes failed to return that evening. Nobile worried that their smoke signals and flares were insufficient, so he and Biagi fashioned reflectors from tinfoil that had been used to wrap their scientific instruments. When airplanes came again, they would light the smudge fires, discharge flares, and hold up the reflectors. He prayed this would work because their fuel supply was almost gone. To make them even more visible, Nobile had instructed his men to soak rags in the remainder of the red dye and lay those out as crosses on the gleaming white ice, marking a possible landing zone.

The next day another airplane appeared in the sky, coming from the direction of Foyn Island. This one was larger than the previous two, and it traced long diagonals across the sky as it searched. It came within a couple of miles, arcing back and forth. The pilot was obviously scouring the ice below, but again, despite smoke signals, flares, and now the use of reflectors, the pilot turned away from them and disappeared. Nobile feared they might never be spotted from the air. "I had known for some time how hard it was to find us amid the pack," he wrote in his journal, "but in practice it seemed more difficult than I had expected. Even Behounek, who up till then had shown great faith that they would reach us, began to doubt it." Viglieri inventoried the stores and reported that they were down to the last scraps of polar bear meat, so they would soon be back on a strict ration of pemmican and chocolate.

Early on the morning of June 20, Biagi reported that he had just received a message from the *Città di Milano* that Italian pilot Maddalena had started from Kings Bay in a large Savoia-Marchetti S-55 hydroplane. Nobile knew the massive twin-hulled aircraft well: just two years before, the same airplane had set a dozen world records for distance, speed, and altitude flight with payload. The Italians were justifiably proud of this magnificent machine. Nobile knew exactly what to look for in the sky, and he told everyone that the minute they spotted it, to light the smoke signals. Biagi should be at the wireless, transmitting and trying to make contact directly with the airplane. As everyone took their positions, Nobile and Cecioni dragged themselves outside the tent and propped themselves into seated positions against some gear.

After two hours, Nobile heard the first faint buzzing of engines, and moments later the airplane appeared on the horizon. Biagi watched it intently, sending precise directions: "The tent is on your course, less than two miles in front . . . go straight ahead . . ." When the airplane responded, obeying Biagi's instructions, they knew it had heard them. It swooped downward to about one hundred yards above the pack and came straight toward them, and then it was upon them. "The throb of the engines grew louder," Nobile recalled of the moment, "and now we could clearly see the [Italian] colors painted on the wings. It was very close."

The airplane roared over them and passed into the distance until it was just a fleck on the skyline, and then they could see it growing larger, coming back. As it came over, they saw someone leaning out of the cockpit, tossing boxes and bags. It made one pass, turned around, came back, and dropped more parcels as Nobile and the others tried to mark where the packages had fallen. After the second pass, the S-55 kept going, returning back to base at Kings Bay.

Viglieri, Biagi, and Trojani started over the ice to locate and collect the dropped items. It was quite a hunt. Some bags had landed between ice boulders and were covered with snow; others had lodged down into crevasses. When the men had scoured everywhere, Nobile did an inventory at the Red Tent. There was some tinned food, two sleeping bags, six pairs of winter boots, more flare guns, and two inflatable rafts. One package contained two rifles, which had broken when they hit the ice, and extra batteries that had also shattered.

Everyone was elated. The sleeping bags, boots, and food were the most welcome, plus the knowledge that they had been located once. Surely there would be constant sorties coming back. But Nobile also clearly understood the logistical challenges of getting them off the ice. The large S-55 could make drops, but it was unsuited to landing on shattering polar pack—only smaller aircraft could attempt such a landing, and successfully taking off again afterward was no guarantee.

Nobile wrote a lengthy message for Biagi to transmit. In it, he requested another immediate drop, taking advantage of the clear skies and calm air before any fog set in and made them difficult or even impossible to spot. He told them to hurry, as ice was deteriorating around them. He asked for more batteries—better protected this time—and medical supplies, including splints for broken bones. Nobile closed his message with the following request: "Please report news of Malmgren's party with the least possible delay. We are anxious about them." Nobile had no way of knowing that five days before, Mariano and Zappi had walked away from Finn Malmgren, leaving him for dead on the ice.

The "immediate drop" Nobile had requested came two days later. This time, two airplanes arrived and made several passes each, dropping scores of packages and crates. Most of them had been attached to small red parachutes, and this time they landed softly, protecting their contents. It was "manna from heaven" to Nobile and his men. That evening, for the first time in nearly a month, instead of pemmican and bear meat they enjoyed tinned beef followed by sweet cakes, and they chain-smoked cigarettes from the cartons that had been delivered. They helped Cecioni re-splint his right leg with a proper medical splint, and he and Nobile were able to take pain medication for the first time since the deadly airship crash.

On the evening of June 22, two Swedish airplanes—they came low enough that their national colors were clearly identifiable—buzzed overhead, dropping a half-dozen parcels. On one of packages, written on the brown paper wrapping, was the following message: "If you can find a landing ground for aeroplanes fitted with skis (min. 250 meters), arrange red parachutes in T-shape on the leeward side." Trojani and Viglieri immediately gathered all of the red parachutes and went a couple hundred yards to the southwest of their crumbling "Bear Floe," where they found a flat section of ice with no obvious cracks or ridges, and just a few slight knolls. It was about 325 yards long and 250 yards wide, with no water leads running through it—at least not at the moment. That could change instantly, but for now they laid the parachutes out in the T-shape as instructed, weighting the bright red silk chutes down with blocks of ice.

That night, with the expectation of an arriving ski plane, Nobile was restless. He knew that any aircraft capable of landing on the short ice runway would have limited passenger space and must likely therefore take only one or maybe two of them off the ice at a time. They would have to make at least three—and possibly six—successful landings and takeoffs to fly everyone away to safety, and with uncertain weather and deteriorating pack, this would be nothing short of miraculous. He doubted the weather and ice conditions would hold out but hoped for the best outcome.

Biagi was having trouble with the radio. For days he had received no new messages or reports, and the handwritten message on the parcel tossed from the Swedish airplane proved that something was not working properly. As he tinkered with the transmitter, Nobile had the others load all of the food, the stove, and emergency equipment into the inflatable rafts. If the pack broke up significantly, they might have to take to the water, and at least they would have enough food to survive for a time as they tried to row their way toward land.

That night after their meal, they heard the now distinct and welcome sound of two airplanes humming their way. Everyone exited the tent, with Nobile and Cecioni, as usual, the last to painfully drag themselves out. Circling around above them were a seaplane, and a plane fitted with skis. The seaplane remained high, and the smaller aircraft, which Nobile recognized as a Fokker military plane, flew in low, made one pass, and descended into the wind, now on an obvious trajectory for landing. Nobile saw it disappear behind masses of ice, then rise from the pack and speed ahead. It looked as though the pilot had been testing the ice with a practice landing, and indeed, he curled back around, and they watched as the airplane came in low and slow, touching down and skittering along, hopping over a few ice bumps and coming to a stop. The Fokker turned around, taxied closer to the Red Tent, and then halted well before a large water lead. The pilot got out and walked toward the tent, while the copilot remained in the airplane with the engine still running.

Biagi and Viglieri met the pilot. He was Swedish military aviator Einar Lundborg, a strapping man with a rugged face and sparkling blue eyes. They guided him the rest of the way to camp. Lundborg followed them carefully as they wove in a circuitous route across narrow water leads to the Red Tent. There, amid the squalor of equipment, inflated rafts filled with sacks and gear, sat Umberto Nobile, propped up against a duffel bag. Lundborg recognized him from many photographs he had seen in the papers, but noticed his long, scraggly beard and moustache; in all the pictures he had ever seen Nobile was clean-shaven. One of his feet wore a civilian shoe, the other, shoeless, wore a stocking and reindeer-skin slipper, and above it was a splint wound with drab bandages.

Lundborg saluted Nobile. "I am Lieutenant Lundborg of the Swedish Air Force," he said in English. Nobile's men helped him to stand on one leg, and the general reached out both arms and drew the Swede in with a firm embrace. Then the others eased Nobile back down. Nobile thanked Lundborg profusely for all of them, commending him on his daring ice landing.

"General, I have come to fetch you all," Lundborg announced. "The landing ground is excellent, and I shall be able to take away the lot of you during the night. But you must come first." Nobile protested, pointing to Cecioni, who was busy trying to rig a litter out of the remaining parts of the crushed *Italia* control car. "Take him first . . . That is what I have decided."

Lundborg shook his head. "No, I have orders to bring you first. We need your instructions to start looking for the others." By this he meant

Malmgren and the shore party as well as the six men last seen floating away in the mortally wounded *Italia*. There was an awkward silence, disturbed only by Titina's excited yipping, which had been incessant since the stranger had arrived out of the sky. Nobile had done some thinking about the order of the men to be taken. He had settled on Cecioni first, since he was badly injured; Behounek next; then Trojani, who had been suffering from a fever during the last few days and could use medical attention; Nobile would be fourth, followed by Viglieri, since he could use a sextant to continue fixing their position, and finally Biagi, who would still potentially be able to communicate by radio. Nobile's initial thought was that he should be last. It was an important tradition and symbol, the captain being last to abandon his ship. But that had made little practical sense, as he was incapacitated and unable to function on his own.

"Please take him first," Nobile appealed again, nodding at Cecioni, "That is my decision."

But Lundborg was firm. He pointed to the airplane, its propeller still whirring. "We will take you to our base not far from here; then I can come back for the others . . . please come quickly." Nobile looked to his men. They discussed it in Italian, and Viglieri, Trojani, and even Cecioni—who had tears in his eyes—implored Nobile to go first. Cecioni added, rather dramatically, "Then, whatever happens, there will be somebody to look after our families."

Nobile reluctantly agreed, and he crawled back into the tent to gather up a few of his personal items. Lundborg stood regarding the encampment, amazed that the men had been surviving here for over a month. "The tent," he observed, "about three yards square, was surrounded by a mass of rubbish . . . tin cans . . . furs, boxes, remnants of the gondola, charts, various instruments, bits of piping, flags, and, a little to the side, the skeleton of Malmgren's polar bear in a large water pool with brownish red spots and remains of the entrails . . . and a large paw . . . exactly where the bear had been killed." A part of the bear's haunch, and strips of meat, were lain over an aluminum pipe from the airship gondola, drying in the sun.

Nobile crawled back out of the tent. He embraced his comrades, kissing each in turn on their cheeks and encouraging them to remain strong. Nobile put Viglieri in charge, telling him to get Cecioni ready to go next. "And whoever comes last, don't forget the little picture of the Madonna." The image remained tacked to the tent pole. Lundborg found the good-byes moving. "All the tears shed out there," he wrote of the departure, ". . . were not tears of sorrow, but . . . of joy because the first stage of saving the whole group was about to begin." Then Viglieri and Biagi lifted Nobile as gently as they could, one by his torso and the other by his legs, and carried him toward the Fokker. They struggled slowly across the uneven surface, trying not to jostle his broken leg. Halfway there the copilot came to assist, and the three of them managed to heft Umberto Nobile into the rear seat, just behind the copilot. Titina, who had followed along, was placed on a tarp at the foot of the passenger seat, beneath Nobile's legs.

Nobile waved once more to Biagi and Viglieri and they started back toward the tent. Lundborg spun the Fokker into the wind, revved the 450 horsepower motor, and sped along the snow and ice, bumping and skipping toward a looming ridge until the airplane lifted into the burning midnight sun, banked a long slow turn over the Red Tent, and winged away west.

42

UPENDED

UMBERTO NOBILE SAT IN A HALF STUPOR ON A GRAVEL BAR. HE WAS shivering and his leg throbbed with pain from being jostled and banged as the Swedish airmen had lifted him from the Fokker. Lundborg tried to make him comfortable by wrapping him in a heavy blanket and wriggling a fur-lined sleeping bag over his legs. Nobile looked around at the Swedish Advance Base—which was a just a long, snow-covered spit on Ross Island (Rossøya), the northernmost island of the Spitsbergen archipelago.

The "Swedish Advance Base" was comprised of two seaplanes and Lundborg's Fokker out on the ice. A big driftwood fire was burning nearby, and a cook pot was nestled into the embers. The airmen gave Nobile some warm soup and coffee. Lundborg told Nobile that in a few hours he would begin making flights back to the Red Tent to pick up survivors, probably one at a time, perhaps two. It was a short flight—just over an hour. At that point, another airman would transport Nobile to the *Città di Milano* at Danes Bay in one of the seaplanes.

Lundborg said that a large ship, the *Quest*, would arrive at Ross Island later. Using it as a communications base and command center, they would continue in their searches for Malmgren, Mariano, and Zappi, who had yet to be located, as well as for any possible survivors from the *Italia*'s envelope, those who had drifted away. He appreciated that Nobile had agreed to leave the Red Tent, because once Nobile was safely aboard the *Città*, he could help direct the searches from there. But first, they needed to get a message out. Lundborg was able to send it by radio from one of the airplanes out on the ice, where it would make its way around the world. It was succinct: GENERAL SAVED. RESCUE WORK CONTINUES.

"I'm off," Lundborg announced. Nobile roused from a dream, lying by the warmth of the driftwood fire. He shook Lundborg's hand and watched him stride away toward the Fokker. Nobile pictured his men at camp anxiously awaiting Lundborg's return. He prayed he would be reunited with them all soon.

Einar Lundborg was flying alone this time. He wanted his plane to be as light as possible. After retrieving Cecioni, who was a large man, he believed he might be able to reduce his trips by fetching two men at a time. He had left everything at base camp—his rucksack, tent, sleeping bag, cookstove—and brought only his rifle and pistol. It was risky leaving his emergency kit behind, but weight would be a concern with every takeoff from the ice—he had barely cleared that big ice ridge in leaving with Nobile.

Lundborg flew low as he neared Foyn Island, flanked by a seaplane escort. He scanned below for any sign of Malmgren, Mariano, and Zappi, but saw only seals sunning on the ice and a few polar bear tracks. Between Broch and Foyn Islands his engine started to sputter. He shut the engine down and glided for a few seconds, testing the gasoline tanks one by one, but the engine kept coughing and cutting out regardless of which tank he used. He deduced that there must be a fuel line problem, or perhaps water in the gasoline. He started looking for reasonable places to make an emergency landing, then remembered that he had none of his gear. Better to try to push on; it was only eighteen or twenty miles to the Red Tent.

And then, there it was. He could see the tent and camp clearly. Two men stood near the radio aerial, and one was lying down—it must be Cecioni. Smoke rose from their signal fires and Lundborg dropped down, realizing he would have to land in a side wind to take advantage of the entire length of the landing zone. The red parachutes were whipping sideways in the gusts. There was no chance for a test landing. Instead, he brought the Fokker directly down, the skis hitting the ice with a jolt. He slid and bounced along, pitching to one side, then felt the ski tip on the right side dig in and submerge in melted snow, and sensed the tail of the Fokker rising behind him as the propeller now plowed into the snow. Then the world spun as the Fokker front-flipped, landing and skidding, with snow and ice pouring into the cockpit. Lundborg's face bashed into the instrument panel as the airplane ground to a halt. Lundborg hung by his shoulder straps, upside down.

He worked to unbuckle the straps and fell onto his head and shoulders, then crawled out of the cockpit onto the ice. He stood and watched the seaplane circle around, then walked away from the Fokker and waved his arms to show that he was uninjured, though blood was streaming from his nose. The seaplane turned west, headed back to base to report the crash. Lundborg assessed the overturned Fokker. One wing strut was bent; the other broken, and a large piece was missing from one propeller. Even if he and the survivors could manage to flip it back over, the Fokker was too damaged to fly. He turned and could see Viglieri, Biagi, and Behounek standing next to Cecioni, whom they had carried to the side of the landing field in preparation for sending him off. They all stared dumbfounded at the upended, shattered airplane. Now, they were six men again on the ice at the Red Tent, and Lundborg would need to be rescued too.

Nobile stirred when he heard a seaplane coming. He sat up and watched it splash down gently on the open water just offshore. A little while later he saw a Swedish airman tramping up the gravel bar. "The Fokker has overturned," the man said, "but Lundborg is unhurt." Nobile shut his eyes in despair. Disaster heaped on disaster. Would it never end?

Nobile considered the dilemma, then asked what other ski planes were at their disposal. There was a Finnish plane that could be adapted with skis, but that would take some time and they did not have all the necessary tools and equipment there at the remote island spit. It might be possible once the *Quest* arrived. The best plan, the Swedes said, would be to have the Italian government request a few De Havilland Moths from England. They were small, light, and maneuverable airplanes; when fitted with skis they were ideally suited to snow and ice landings. It would take time, however, perhaps weeks, to ship them to Spitsbergen.

Nobile winced with both pain and worry. He later admitted to "a sadness in [his] soul" as he thought about the men marooned at the Red Tent. He was safe, delivered, while they remained in the churning maelstrom of ice cakes and floes. Nobile finally looked up and asked whether the Swedish rescuers had any spare pemmican they might drop them. They said they would do so that very evening. That settled, there was nothing left for Nobile to do but get to the *Città di Milano,* and he requested transfer immediately. Soon the Swedes had carried him out to a seaplane, and he was airlifted directly to Virgo Bay.

Nobile lay in his cabin aboard the *Città di Milano,* staring at himself in the mirror. He was shaky and feverish. "I looked frightful, unrecognizable," he wrote of seeing himself for the first time in over a month, "with a long, bristly greyish beard smothering my face." Splotches of grime covered his forehead and neck. His cheeks were windburned, his lips split, and he appeared gaunt from his thirty-two days on the ice. His clothes were rank and putrid, and he turned from the mirror, sickened by his own stench, and his appearance, which reminded him of the horrible crash. But he was also heartsick from his recent visit with Captain Romagna. When Nobile had been lifted onto the *Città di Milano,* the Italian crewmen had cheered and shouted with joy at his safe return. But as a corps of international journalists took pictures and barked out questions, Nobile was quickly whisked to a cabin belowdecks and instructed not to speak to the press, as Mussolini's Fascist government intended to carefully manage the messaging. A little later, Captain Romagna had come below. Nobile informed him that he was ready to take command of the rescue operations and begin directing them immediately.

Captain Romagna was curt in his response. Orders from Rome, he said, coming directly from Prime Minister Benito Mussolini and Italo Balbo, Undersecretary of the Air Force—were clear: Captain Romagna himself would remain in charge of communications and rescue operations from the *Città di Milano*. As for Nobile, he needed to see the ship's doctor to properly attend to his broken leg and try to regain some of his strength. Then Captain Romagna turned to leave, but before departing, he said, "People might criticize you for coming first, General. It would be as well for you to give some explanations."

Nobile was simultaneously confused, hurt, and stunned. Explanations? The pilot Einar Lundborg had practically forced Nobile to be the first man taken from the ice, explaining that he was needed to supervise rescue operations. Captain Romagna just shook his head. He said that he had issued no orders to Lundborg. Any such directives must have come from the Swedes. Nobile was incensed. "However that may be," he said in disgust, "the fact is that I did not leave by personal preference. I am here only because I was told it was of great importance that searchers use my directions to find Mariano and his companions, and for the six who disappeared on the bag of the *Italia*."

Romagna just stared blankly at Nobile. He reiterated that the general would need to prepare an explanation for leaving his men behind, and then he left the cabin. Nobile fumed. He felt betrayed by his own government, and also felt under no obligation to explain himself to anyone— least of all his own men, who knew he had urged, nearly begged Lundborg to take Cecioni first. But he would deal with that mess later. For now, all he could think of was his men at the Red Tent, and what he might be able to do for them. He managed to convey the following message to the Swedish Air Force, to be sent to Biagi: "Don't be anxious. I am here. The Finnish trimotor is being fitted with skis . . . we have ordered 2 or 3 small planes from England. You will receive 6 more batteries from the Swedes, some smoke signals, a tent, various medicines, dried milk, 110 lb of pemmican . . . I am reckoning on seeing you all again very soon . . . keep your spirits up . . . Your Nobile."

With that taken care of, Nobile decided that he would do all he could to ensure that his men were rescued. He learned from his wireless operator Pedretti—and from a few of the crewmen aboard the ship—that the Italians themselves had been slow and ineffective in their approach to the rescue. They were anything but a coordinated central command. Rather, the multinational effort was taking place haphazardly, with various countries essentially freelancing their individual efforts. Nobile also learned that Roald Amundsen had flown from Tromsø in a French plane to rescue the general but had not been heard from since. Nobile could hardly fathom the cruel irony: Roald Amundsen had courageously come to save him, only to be lost himself. It was hard to believe, but it was true. Now, they said, some of the Norwegian pilots, including Riiser-Larsen, were being diverted from their sorties looking for Mariano, Zappi, and Malmgren to go search for Amundsen. And there was more. An American woman named Louise Arner Boyd, an immensely wealthy socialite who was also an Arctic explorer, had chartered the ship *Hobby* and would be financing—as a gesture to the Norwegian government and the people of Norway—an extended search for Amundsen.

It was all a great deal to take in. Umberto Nobile was overwhelmed by grief at the state of affairs, and it was difficult not to blame himself for the crash of the *Italia* and the series of disasters the accident had set into motion. As the airship's commander, all blame—warranted or not—must land on him. But there would be plenty of time to process the guilt and explain to the world everything that had led to the *Italia* falling from the sky. At the moment, his thoughts were dominated by getting the survivors safely from the Red Tent. That seemed well enough under control: the Swedes and Finns would work tirelessly toward that goal. But what haunted Nobile the most was the fate of Mariano, Zappi, and Malmgren. He had last seen them walking away from the Red Tent more than a month ago, and there had been no sign of them since, despite numerous airplanes searching for them for the last few weeks.

Nobile lay back down in excruciating pain, waiting for the ship's doctor to come check on his condition. He tossed and turned, fever burning his forehead, and conjured the image of Mariano, Zappi, and

Malmgren somewhere out on the pack. He knew they would be dangerously low on food by now, and should they encounter polar bears, they would be in desperate trouble, for their only weapons were their knives and an axe.

43

HELP

MARIANO CLUTCHED AT HIS EYES AND SCREAMED IN AGONY. THE pain came in searing waves, with pressure pushing from deep inside until it felt like his eyeballs would burst, coupled with the sensation of knife blades slicing across his eyes. He buckled to his knees on the ice, holding his hands over his face and crying out. Zappi hunched beside him. He tore a long strip of tattered cloth and wrapped it around Mariano's head, covering his eyes with it.

Mariano was snow-blind. He had felt it coming on a few days after they had abandoned Malmgren. He and Zappi had only the one pair of snow goggles between them, and though they had shared them, the constant reflection of the sun's ultraviolet rays off the snow and ice had damaged his corneas. Zappi cinched the bandage tight across the back of Mariano's head and told him not to rub his eyes—it would only make it worse. But the penetrating pain was accompanied by intense itching, too, a feeling like his eyes were filled with sand, and it was nearly impossible to keep from rubbing at them with the back of his hands. As he reached up to tear away the blindfold and scratch at his eyeballs, Zappi grabbed his arms and pinned them behind his back, warning him that he must let the sensation pass. If he did so, his eyes would likely improve in a day or two.

In the meantime, Zappi would now have to lead Mariano over the ice. For many days they had been walking toward what they believed was Foyn Island, but after each twelve-to-fifteen-hour march, when they had surveyed the skyline, the island appeared no closer, and sometimes it even looked farther away. No matter how long they trekked, their drift erased every step of progress. Now, with Mariano reduced to hanging onto Zappi's arm or shoulder, their progress would be even slower. And even more concerning, they were running out of food. They had started from the Red Tent with about two week's rations for three men. Because they had left some food with Malmgren, in case he miraculously revived, they were now down to just a few days' rations of pemmican.

Zappi knew that they must reach Foyn Island at all costs, for there he could at least forage for shorebird eggs, or possibly sneak up on a nesting bird and kill it. He knew that guillemots also nested in large numbers in the island cliffs, and perhaps he could summon enough strength to climb up and capture some. Beyond being spotted and rescued, it seemed their only hope.

Mariano was crying out less often; the blindfold seemed to be working, giving his eyes a few days' reprieve from the omnipresent glare. But because he could not see where he was going, he constantly sloshed through knee-deep pools that Zappi managed to avoid, soaking his feet. The summer ice breakup was compounding the problem, creating wider and more frequent leads. Once, Mariano fell in the water up to his waist. Zappi acted quickly, producing the rope they had brought with them, and he managed to haul Mariano out before he submerged completely. But that night, Mariano told Zappi that he could tell that both of his feet were badly frostbitten. They remembered Malmgren's blackened and necrotic foot, and they knew the grim truth of the situation: in a day or two, Mariano would be unable to walk, and without medical treatment, he would succumb to gangrene and die.

So, they trudged on toward land. Zappi told Mariano that he could see the outline of cliffs more clearly now; surely it was only a day's walk, if only the drift would ease up, or the wind shift. Then one day—it must have been June 21 or 22—Mariano stopped abruptly, held up his hand, and whispered, "I hear a plane." Zappi halted and listened for a minute, then dismissed the sound as only the wind whining over the pack ice. But Mariano said he was certain, and Zappi cupped his hand to his ear and craned to hear—and there it was, just a faint buzzing. He looked to the sky and way off in the distance, sure enough, was a tiny black dot moving slowly across the gray-white horizon.

Zappi dropped to his knees and threw off his rucksack and rooted frantically inside. "The matches! We must light a signal fire!" The sound of the airplane was getting louder as it drew closer, and Zappi found the matches and tore a few strips of cloth, knotting them with his bare fist into a tight bundle. He poured some fuel from a small canister onto the rags and with trembling hands, struck a match. It broke off short, and he pinched another from the match box, managing to strike this one afire. He touched it to the rags, and they burst into a small flame, then sputtered out in the wind. He tried again and again, pouring out more of their precious fuel and lighting match after match, but the bundle merely flamed for a moment, then extinguished. Zappi looked up in anguish as he watched the airplane banking away toward the east. They stood and waved their hats, jumping up and down, but it was no use. The pilot had not seen them.

Though overcome with bitterness, Mariano tried to remain positive. If one airplane was out looking, surely there would be others. They must keep the faith. Zappi, though distraught, agreed. Faith was all they had left. They must hope and pray that there would be others, and that one would spot them. Given Mariano's feet and the fact that they were almost out of food, it appeared their only chance at survival.

Over the next few days, Mariano's eyes improved enough that he could remove the blindfold for short periods. This helped both his morale and his mobility, for during those times he could let go of Zappi and stumble along on his own, weaving through upthrust ice pinnacles and stepping across leads. The condition of the ice was deteriorating badly, and it fractured in all directions, grinding and roaring where one floe pressed against another. Occasionally they found freshwater that had leached out onto the tops of floes and pooled in low clefts, and they dropped to their hands and knees to drink and fill their canteen. The water, at once sweet and salty, burned their raw lips as they lapped and slurped it from the surface like sled dogs.

One day, as they snaked slowly across a field of dense hummock ice, Mariano saw the outline of land on the horizon—it must still be either Foyn or Broch Island—but it now appeared behind them. He pointed it out to Zappi, who simply nodded and kept walking away from the dim profile, looking at his compass. If they had drifted beyond that pair of small islands, perhaps they were nearing North East Land. Sometime later, a pair of airplanes appeared, dark specks far off in the sky, heading away from them. They surmised that the airplanes must be in the vicinity of the Red Tent, and this gave them a moment of contentment: at least some of their party would be rescued.

Mariano's eyes began to burn again. He had gone too long without the blindfold, so he pulled it back over his eyes, took Zappi's wrist, and followed, lurching along behind him. Suddenly, Mariano blundered over a sharp protrusion of ice, felt his leg slip into a crevasse, and fell awkwardly, with all of his weight coming down on one side as his right knee wrenched and twisted. He yelped out, then lay writhing on the ice and clutching his knee with both hands. Zappi knelt next to him. Through clenched teeth Mariano said he thought his leg might be broken.

Zappi took out his knife and cut Mariano's right trouser leg from the ankle up to the knee. The flesh of the leg was swollen, dark and puffy and putrefied, and deeply frostbitten, much as Malmgren's had been. But as Zappi felt gently up and down the leg, it seemed unbroken. Likely he had sprained his knee. After a time, Mariano said the pain had subsided. He would try to go on. Zappi wrapped Mariano's torso in his arms and heaved the big man to his feet, but the leg would not bear weight and he crumpled back down in a heap.

"Save yourself," Mariano whispered to Zappi. Zappi looked out across the expanse of crenelated ice. He knew that Mariano could go no further: he was half snow-blind, with severe frostbite and now a useless leg. But in his own weakened condition, he would not get far on his own either. By now he figured it was the end of June, a month since they had left the Red Tent. Better to remain with Mariano. He already had leaving Malmgren on his conscience. Zappi dragged Mariano next to a large ice boulder that served as a windbreak. Using the axe, he feebly chipped off some ice blocks and built a low wall extending from the ice boulder, making a semi-circular shelter. He took every piece of clothing from both rucksacks and bundled Mariano in these, covering his legs as best he could. Then he gave Mariano his last chunk of pemmican. It was here, Zappi realized, where they must remain, either to be rescued, or to die.

Zappi fumbled in the bottom of his rucksack and took out the last of the colored strips of rags he had hoped to use to light signal fires. He walked some distance away from the makeshift shelter and found a relatively flat section of ice. Taking long strips of the fabric, he began laying pieces down onto the ice, holding them in place with fist-sized ice balls. He walked forward and backward, side-to-side, spelling out words in cloth until he was finished. Maybe, by some miracle, an airplane would see it:

HELP FOOD ZAPPI MARIANO

44

COMRADES

UMBERTO NOBILE DRIFTED IN AND OUT OF FEVER DREAMS. THE doctor had given him strong medicine for his pain, and he dozed and woke intermittently. There was a lot of commotion outside his cabin door, men coming and going, muffled voices in Italian, Norwegian, English, German, Swedish. Some, he could tell by their questions, were reporters. He strained to hear but drifted back off into hazy dreams: He was aboard the *Italia*, the men at work at their posts; the *Italia* plunging downward, descending into a realm of ice and sky. Then the violent impact, and he would awaken, shaking and sweating.

On July 2, Nobile roused to the sound of a familiar voice. It was Hjalmar Riiser-Larsen. Nobile drew himself upright, wresting his mind from yet another nightmare. "I was in bed with fever," he wrote of the visit by Riiser-Larsen, "partly physical but far more the fever of my sleepless brain perpetually reverting to the men in the tent and the others who were lost." As Nobile strained to focus on the large form looming over his bed, Riiser-Larsen bent to embrace the general. Riiser-Larsen had been instrumental in helping to pilot and navigate the *Norge,* and despite the post-flight unpleasantness, the two men shared a deep respect for each other. There was an unspoken bond between them, the connection forged by their shared achievement in being first to fly over the North Pole.

For a long time neither man spoke. Both men's eyes filled with tears as they regarded each other. In Nobile, Riiser-Larsen saw a broken man—a leader tormented by the loss of lives under his charge. Gaunt and pallid, Nobile seemed to have aged well beyond his forty-two years. His dark eyes bore a far-off sadness. In the tall, strapping, square-faced Norwegian, Nobile saw a man in his prime, not yet forty years old, vigorous and purposeful. He had come to pay his respects to Nobile before setting off in search of his friend and countryman Roald Amundsen.

For the passionate, sensitive Nobile, the gesture of his coming prompted a flood of memories and emotions. Nobile's thoughts raced with everything he wished to say. "I wanted to thank him," remembered Nobile of those long, wordless moments, "to say that I felt an indissoluble bond united all who had shared the *Norge* flight. I wanted to tell him my unhappiness at the misunderstandings that had arisen; all my sorrow for what Amundsen had said about me, and my regret at having to defend myself. I wanted to express my gratitude for the chivalrous gesture with which Amundsen and he had wiped out those miserable little squabbles of the past . . . this was what I longed to say—what I wanted him to read through my tears. But I was silent. I could not speak."

He did not have to speak, for Riiser-Larsen was able to read through Nobile's tears. He leaned down and embraced Nobile, wrapping him in his giant arms and pressing him close, an embrace that said he understood, that there were no hard feelings, that their bond was, after all, unbreakable. They were comrades.

When they finally spoke, Riiser-Larsen tried to comfort Nobile by telling him he must have faith, that his men at the Red Tent would be successfully brought off the ice. Then he asked Nobile about Malmgren, Zappi, and Mariano. He wanted to know what provisions they had taken with them. Riiser-Larsen was particularly fond of Malmgren, with whom he had also bonded on the *Norge*. Nobile said they had taken about 120 pounds of pemmican. "Oh, that's all right then," Riiser-Larsen said optimistically. "You needn't worry about them. They have enough for forty-five days." This bolstered Nobile's spirits, and the talk turned to Amundsen.

Riiser-Larsen was rightfully concerned. It had been two weeks since Amundsen and the French flight crew had left Tromsø, and since that last cryptic wireless message—DO NOT STOP LISTENING. MESSAGE FORTHCOMING—not a word had come. Global headlines expressed the anxiety everyone was feeling, but Riiser-Larsen tried to remain positive. Having spent nearly a month on the ice with Amundsen in 1925, he knew the man's toughness, his guile, his uncanny ability to survive even the direst of circumstances. In the first few days after Amundsen failed to arrive at Kings Bay, many speculated that Amundsen, without telling anyone, had bypassed Kings Bay entirely and flown directly toward the area of the *Italia* wreck. That would be just like him. He had on numerous occasions before kept his intentions secret and changed plans at the last minute. The most famous of these was in 1909, when, having learned of Peary's conquest of the North Pole, he secretly altered his own North Pole plan and headed instead for the South Pole. And the greatest of all the explorers had been presumed dead countless times before, only to rise from the polar pack, his steel-gray eyes flickering with a glint of mischievousness.

But now, after two weeks, and still no word from Amundsen, the world was worried. Had some disaster finally befallen "The Last Viking," one even he could not escape? Riiser-Larsen held out hope, and he told Nobile that he was confident he would be found or return on his own as he always had. In any event, Riiser-Larsen explained to Nobile that his government had called him off the search for Malmgren and the others. He would be leaving shortly to look for Amundsen, first by airplane, and then, if that failed, aboard the *Hobby,* which he said had just been charted by American Louise Arner Boyd. Riiser-Larsen would direct the search from the ship.

Riiser-Larsen apologized that he must abandon the search for Nobile's men, but orders were orders. He pointed out that the search would continue, with numerous countries still involved. Of note, the icebreaker *Krassin*—the most powerful such vessel in the world—had just reached the pack ice and was churning its way along the northern coast of Svalbard, headed toward the survivor's camp at the Red Tent. After voicing these assurances, Riiser-Larsen stood to leave, and the two comrades bade each other goodbye.

Riiser-Larsen's visit had improved Nobile's mood, but Nobile remained helpless to do much himself. He appealed to Captain Romagna to let him join the searches by airplane, and even asked to be flown out to the *Krassin,* where he might assist its captain, but Nobile was told that orders from Rome were that he remain aboard the *Città di Milano.* Nobile soon learned that *carabinieri*—Italian police guard—had been posted outside his door, ostensibly to keep the press from hounding him, but he suspected it was also to keep him from leaving.

It was true that reporters, journalists, and film crews had arrived at Danes Island in droves ever since the crash of the *Italia.* Nobile found it tragically ironic that just a handful of press had been present to see the *Italia* take off on its historic Arctic flights, yet disaster had brought them in droves, circling like sharks to blood. For now, all General Nobile could do was stay in constant contact with Pedretti, his wireless operator aboard the *Città,* and keep apprised of the rescue proceedings. And Nobile must prepare an answer to the question that most of the journalists seemed obsessed with: Had Nobile abandoned his men by being the first one carried from the ice?

45

THE *KRASSIN*

THE STEEL-SHEATHED RUSSIAN ICEBREAKER *KRASSIN* BASHED through rough pack ice, shrouded in thick fog. Three hundred twenty-five feet long, with three coal-fired steam engines producing 10,000 horsepower, the *Krassin* could cleave through ice up to fifteen feet thick and reach a top speed of eleven knots. Of the 110 crew members, over half were engine room personnel tasked with the backbreaking labor of shoveling coal to feed the hungry beast, which consumed some six tons of coal per hour. The *Krassin* had brought onboard a Junkers trimotor airplane fitted with skis, and, in the hopes of locating the *Italia* survivors, they had brought ten complete sets of thermal underwear and clothing. All told, there were 138 persons on board, including a handful of Russian reporters and Italian journalist Davide Giudici[1], who had managed to finagle passage as the ship had been coaling at Bergen. The ship's skipper, Captain Karl Eggi—a burly Estonian—happened to be an old friend of Umberto Nobile's.

The *Krassin* had worked hard in making it to just west of Rossøya, where the Swedish Air Force had its remote northern base. The ice had been so dense that the ship had become beset for three days, unable to move. When winds and currents finally freed some of the pack, the *Krassin*

---

1 The industrious Davide Giudici was the very journalist who had gone to Amundsen's home and interviewed him prior to Amundsen leaving to go search for Nobile. He immediately wrote a book about his experiences called *The Tragedy of the Italia: With the Rescuers to the Red Tent* (London: Ernst Benn LTD, 1928).

crawled slowly toward the northeast at just a mile per hour, crushing condensed ice as it went, using its power and bulk to carve an open lane for itself. But at times the ice was so thick and impenetrable that the captain would have to back away, then charge forward at full speed and ram the floe, the impact shuddering through the ship like a tremor, sometimes toppling crew members on the deck and in the engine room. The constant grating and tearing of the ship's hull against the ice created a terrible sound inside the ship, bringing some crewmen to cover their ears, begging for the horrific sounds to stop.

The journalist Giudici was impressed by the spectacle of the *Krassin's* battle against nature. "For hundreds of yards around," he wrote, the pack ice crumbled beneath the formidable blows of the ice breaker. Great hummocks of fantastic shapes were pushed up like feathers, afterward overturning and disappearing in the sea amidst columns of white foam. The white expanse seemed to be in convulsion as from the effects of a frightful earthquake."

When ramming failed to pulverize the ice, the captain simply shifted tank ballast astern, allowing the *Krassin* to beach herself on top of the ice like a gigantic walrus coming up to sun, and the weight of ten thousand tons would smash the ice downward, creating a lane. In this arduous way, the *Krassin* plowed to about ten miles north of Kapp Platen, just thirty miles west of Broch and Foyn Islands, where Nobile believed Malmgren, Mariano, and Zappi might have reached. At Kapp Platen, the *Krassin* anchored, and crewmen and mechanics began offloading the Junkers airplane, which would begin reconnaissance flights the next day, July 7, 1928.

Zappi and Mariano lay in the small ice shelter Zappi had dug when Mariano could not continue on foot. They had been there for ten days, completely out of food. Periodically Zappi had risen to go find meltwater leached out onto the tops of ice hummocks, and he scooped what he could into his canteen and brought it back to Mariano, who was now delirious. For the last few days, he had been trying to rip his clothes off in violent fits, and Zappi would have to calm him down, telling him he must remain clothed, or he would freeze to death.

Zappi was also in grim condition, though at least, so far, his feet and legs had not succumbed to frostbite as Mariano's had. He was able to get up and trot around to maintain circulation, sometimes swinging his arms around in windmills, but he was so depleted from lack of nourishment that these bouts of exercise for warmth were exhausting, and he would soon slump back down next to Mariano and pass out. In one of his fever dreams, Zappi received permission from God to return to Rome briefly, but only if Zappi agreed to go back to the ice floes after his visit. In the dream, Zappi raced through many of Rome's finest restaurants, gorging himself on pastas and meats and stuffing a rucksack with food until it was bursting with breads and desserts—these he would take with him, and he and Mariano would have food on their ice floe. But then he was back out on the windswept ice, and when he opened the rucksack, it was empty—all the food had disappeared. After that nightmare, Zappi felt the cold sapping him of his last strength, and he relinquished any further hope, resolved to certain death.

One day Zappi's eyes blinked open. He thought he heard airplanes. He struggled to sit up and scanned the frozen skies. He saw nothing but clouds drifting past. He knew he was hallucinating. It had been happening frequently now. The only positive was that the once terrible hunger pangs had subsided into a dull emptiness in his belly. It had been so long since he had eaten, his body seemed not to crave food anymore.

And then, the next day, Zappi heard it again. A distinct, droning buzz, drawing nearer. He forced himself onto his knees, then stood, nearly buckling back down from dizziness, but he righted himself by leaning against the ice block wall he had built. There was a dark blot against the flat-white sky, and it appeared to have wings. Zappi conjured all of his remaining strength and raised his binoculars to his eyes.

"It's a Junkers," he husked to Mariano.

Mariano had barely moved in many days. More than once, Zappi had assumed he was dead. But now, he stirred slightly, then rolled onto one side and poked his head out from under his hood, squinting up into the sky. He saw it too.

Zappi staggered up to a wind-shorn hummock and climbed on top of it, standing as tall as he could. He pulled off his hat and waved it in the direction of the airplane, but it was already traveling away toward the north. Then it turned in a wide arc and circled back, and as Zappi looked through his field glasses, he could see Russian markings on the fuselage and wings.

"The Russians have come," Zappi cried out, his voice hoarse and feeble. Mariano had risen to his knees, and he saw the airplane passing directly over them. He was certain he could see the pilot waving out of the cockpit as he whooshed past.

"They have seen us," Zappi cried out. "We are saved!"

But the airplane kept on flying, disappearing into the endless vault of sky.

No airplanes came the rest of that day. Or the next. Or the next.

Lying in the ice shelter, they talked it over and were certain it had been an airplane. Surely, they could not have both hallucinated the same thing. That seemed impossible. And yet, for whatever reason, no planes had returned. Zappi wrapped Mariano as tightly as he could with their sodden blanket. He was so frail and wan, Zappi thought, he must have only a day or two left to live. And with that, Zappi curled beside his friend and waited for the end.

A shrill siren blast echoed across the icescape. Zappi was startled awake. He nudged Mariano, who shifted and groaned. Zappi listened intently, rose to his knees, and surveyed the surroundings. Nothing but ice and sky, sky and ice. Light wind blew across the buckled plain and Zappi dropped back down. The siren wail must have been shearing ice somewhere, nothing more. But then, up in the sky, Zappi noticed a column of dark cloud, like smoke curling into the air. He rose again and lurched his way to the highest ice he could find and clambered to the top.

"Mariano," he shouted. "A steamer is coming this way!"

It was the *Krassin*.

46

SALVATION

A CREWMAN STOOD ON THE TOP DECK AT THE RAIL, WAVING AND shouting. He looked through binoculars, saw the dark shape of a man standing on the ice, and what appeared to be another form lying down. The crewman called out, "A man! A man! I can see him!" A few moments later, Captain Eggi blew the ship's whistle loud and long. He repeated the siren blare every few minutes as the *Krassin* rammed through ice more than ten feet thick.

It was the early morning of July 12, 1928. The last two days had been difficult, with slow progress through the pack ice and spotty communications, though some information had been getting through to the ship's wireless operators. On July 10, the Junkers trimotor had spotted two men on the ice—and possibly a third—and made several passes over them. The pilot believed it must be the Malmgren group, though it was impossible to be certain based only on the flyovers. It was also conceivable that it was Amundsen. But soon afterward, the pilot of the Junkers had been forced down in dense fog and made an emergency landing many miles from the base ship *Krassin*. After he went down, the pilot was able to report that what he thought was the Malmgren group had been sighted at 80° 42′ N, 25° 45′ E, on an ice floe surrounded by open water. The pilot sent the following message to the *Krassin*: "*Have landed . . . aircraft damaged but we are safe . . . No one hurt. Food for two weeks. Don't stop for us. Go to Malmgren's aid soonest.*"

With that information, Captain Eggi pressed on toward those coordinates. They could pick up the Junkers and its flight crew later. As they chugged east, the *Krassin* was in contact with the *Città di Milano*, though the communications were intermittent and incomplete. The most pressing news was that Biagi reported deteriorating conditions for the men at the Red Tent: ice leads were opening up all around them, and the size of their floe had been steadily diminishing. They had moved the Red Tent next to Einar Lundborg's overturned Fokker and were using the airplane as additional shelter. Lundborg had been picked up on July 6 by one of the De Havilland Moths and was helping to organize further rescue flights. Biagi said their floe had drifted to within a few miles of Cape Leigh-Smith—they should look for them there, and they should hurry. Some members were now contemplating using the inflatable rafts that had been dropped to paddle toward the mainland, since they were now surrounded by water on all sides.

"There they are!" crewman on the *Krassin* were shouting. Now they could clearly see two men about three hundred yards in the distance, one standing and waving, the other lying down. Captain Eggi ordered the ship's engines powered down; he feared getting too close and disrupting or even upending their floe, which might spill the survivors into open water. He dispatched a small rescue team, led by the ship's surgeon, with ropes, planks, and a stretcher. The rescuers descended from the ship via rope ladders as sailors stood cheering on the bridge.

The ice between the ship and the men was treacherous, with large rifts and open leads. The rescuers used boards thrown down from the deck to span some of the wider gaps, and, leaping narrower ones, navigated their way over to the men. The floe they were on was tiny, not much more than a block of ice. As the rescuers arrived, they approached the man standing. His face was charred dark from exposure to sun and glare, his long beard shaggy and matted.

"Malmgren?" they asked him.

"No," the man answered, his voice little more than a wheeze, "Captain Zappi."

"Where is Malmgren?" they asked. Zappi mumbled something in Italian. He bowed his head, and his shoulders slumped. "Dead," he said at last, his blank eyes welling with tears. He swept his hand in an arc, gesturing somewhere out on the distant ice. "Back there."

The surgeon moved past him and hurried over to Mariano, who lay in a trench half filled with water. The surgeon knelt beside him, placed his hands carefully beneath his head, and lifted it slightly. Mariano opened his eyes; they were distant and glazy, and his body trembled in hypothermic spasms. He managed to force a thin smile, "like that of a man unexpectedly called back to life, whose sufferings were finished." Through quivering lips, he whispered, "Thank you." Then his eyes closed again.

They loaded Mariano onto the stretcher and bore him toward the *Krassin*. Zappi was able to walk on his own. He took one last look at the dugout shelter he had hacked into the ice boulder. Beyond it, on a nearby floe, were the rags and torn garments he had used to write out their SOS: HELP FOOD ZAPPI MARIANO. Then he turned and strode toward the ship, its two tall yellow smokestacks belching plumes of black smoke into the sky.

Once Mariano was aboard the *Krassin*, the surgeon and some crewmen stripped Mariano of his tattered clothing. His pulse was weak and irregular. They massaged him vigorously, then wrapped his torso in warm, dry blankets. His forehead burned with fever, but it was his legs and feet that most concerned the surgeon. The right foot was black and gangrenous; it would need to be amputated. The other leg and both his hands were also frostbitten, but they might possibly be saved. Only time would tell. The surgeon spooned hot broth between Mariano's lips, entreating him to drink slowly. Given Mariano's condition, the surgeon felt sure that had he not been rescued, Mariano would have been dead within the next twelve hours. It was miraculous that either man was still alive: poorly clad and with no shelter or sleeping bags, they had been wandering the ice for forty-three days, the last thirteen of them, they said, without food.

Within an hour of the rescue, the *Krassin* was once more crushing its way through the ice. Captain Eggi sent the following message to General Nobile aboard the *Città di Milano*: *Commander Zappi and Mariano are on board the* Krassin. *We are going towards the Viglieri [Red Tent] group. Please give me the latest position of the tent.*

Nobile was overcome with relief that Zappi and Mariano had been saved, but he could only speculate as to why Malmgren was not with them. With a feeling of dread overcoming him, he did as Captain Eggi had asked and sent the most recent position of the Red Tent. Because the *Krassin* was not able to connect directly with Biagi at the Red Tent, Nobile would remain by the wireless on *Città di Milano*, maintaining contact with Biagi and the *Krassin*, hopefully managing to communicate as an intermediary to guide the icebreaker to the rest of the *Italia* survivors. He sent the following message to Biagi: "*It would be advisable to light smoke-signals. When you see the* Krassin *tell us where you sight it and how far away; then we shall be able to help them find the tent.*"

On the early evening of July 12, sailors on the *Krassin*, scanning with binoculars and telescopes from the deck, began shouting. They could see smoke spiraling into the sky some five miles in the distance. Captain Eggi altered course and bore through ice toward the smoke. It was slow going as the ship's steel-sheathed bow bashed through ice that was in places up to fifteen feet thick.

Around 8:00 P.M., crewmen began cheering. Just a few hundred yards ahead they could see the strange outline of Lundborg's upside-down airplane, its tail raised in the air and its nose pressed down into the ice. Beneath the tail section was the Red Tent. On the ice near the tent, several figures were moving about, waving their arms. Captain Eggi eased the *Krassin* to within one hundred yards of the ice camp then stopped.

The gangway was lowered, and a number of sailors disembarked and strode across the ice. They noticed that multicolored flags had been spread out as a signal to airmen. There were tracks running in all directions in soft snow, and as they drew closer, they could see the radio antenna anchored by guy ropes.

A tall figure emerged to greet the rescuers. It was Viglieri, left in charge when Nobile had departed. Just behind him were Trojani and the big Czech Behounek. The survivors embraced their rescuers, praising them, and God, for their salvation. As Russian crewmen came down the gangway with stretchers, Viglieri led the officers and a handful of journalists over to the Red Tent, where he and the others had survived for seven weeks since the fateful day that the *Italia* crashed.

Cecioni was up and moving around on a pair of crutches he had fashioned from two raft oars. He was laughing and crying tears of joy, overwhelmed with emotion. When sailors arrived with a stretcher, Cecioni said he wished to make his own way, but eventually he yielded to being carried. Now that they had finally been rescued, the survivors seemed in no rush to board the *Krassin*. Instead, they proudly gave the officers and journalists a tour of their ice camp. Standing beside the wireless station was the tireless Biagi, his beard long and thick. He posed for a few photographs, explaining to the journalists how the remote wireless station worked, and how it had been used to contact the outside world. He told the journalists that the last message he had sent, once they'd seen the smokestacks of the *Krassin,* had been to thank the rescuers for finding them. He ended the telegram with the words "Long live the King and Italy."

Nobile had sent instructions to both Biagi and Captain Eggi of the *Krassin* to try salvaging a number of items and equipment from the encampment. He wanted them to bring back all of the scientific instruments, cameras, and the Red Tent itself, which Nobile believed would hold historical significance. With Biagi's assistance, the crewmen helped scour the floe the tent and Fokker were now on; it was a relatively flat rectangular slab of ice about 330 yards long and 220 yards wide, fracturing and cleaving at the edges. During the short search, they found part of the rudder of the *Italia,* which Biagi packed away along with a few rifles which had been airdropped.

Last, Biagi went over to take down the Red Tent, which had been their sanctuary and salvation since May 25. The red dye had faded, and the tent was slightly tattered in places from being whipped by wind, sleet, hail, and snow, but otherwise it remained in good shape. Near the tent entrance, on the wing of the overturned Fokker the survivors had laid out to dry the polar bear rug made from the bear Malmgren had shot; they'd been using it inside the tent as insulation against the freezing ice floor they had been sleeping on for a month and a half. Biagi retrieved his personal belongings—there were not many—and stuffed them in a rucksack. Then, as directed by Nobile, Biagi took from the center tentpole the image of the *Madonna of Loreto,* packed it safely among his things, and headed across the ice to board the *Krassin.*

From the deck rail, Biagi stood and watched as Russian crewmen dismantled the Fokker for transport on the icebreaker. Later, as the *Krassin* weighed anchor, Biagi heard the sounds of "rattling windlasses and creaking capstans," and felt the ponderous ship move beneath his feet. He stared out across the pack, his eyes focused on the deteriorating slab of ice where the tent once stood. He gazed beyond, to the site of the crash, and to the shards of rubble ice being ground and subsumed into the polar sea. Biagi removed his hat and knelt before a pile of gear salvaged from the *Italia.* He bowed his head in prayer, murmuring the name of his companion, the first to die in the disaster. The stern engine car that had served as a makeshift grave had been engulfed by the roiling ice.

"Pomella," he whispered, and the name was carried away by the Arctic wind.

47

HOMECOMING

UMBERTO NOBILE SAT UNCOMFORTABLY IN A CRAMPED RAILWAY car in Narvik, Norway, waiting to proceed by train to Rome. He and the other survivors of the *Italia* crash had been transported to Narvik on the *Città di Milano,* arriving there on July 26, almost two months to the day since the disaster. At the dock, they were met by a few hundred jeering Norwegians, who hissed and booed as Nobile and the other Italians walked along a special wooden gangway leading directly from the ship to two railway cars hired out by the Italian government. The planking had been erected, according to local papers, so that Nobile would not be allowed "to defile Norway's soil by setting his damn feet on it." Public sentiment toward Nobile was vitriolic, bordering on violent. The papers had instructed the crowd to greet the general with catcalls including "Down with Nobile!" and even "Death to Nobile!" The vocal assemblage had obliged, and the men had been quickly ushered to the awaiting rail cars.

Nobile was racked with emotions. He was simultaneously indignant, hurt, confused, contrite, defensive, and deeply saddened. There was much on his mind, and much weighing on his conscience. He peeked out the shuttered window and could see, beyond the hostile gathering, a large poster hanging on the quayside: REWARD—10,000 KRONER FOR ANY INFORMATION LEADING TO THE MISSING EXPLORER ROALD AMUNDSEN. The Norwegians had disliked Nobile stemming from the public squabbles with their national hero after the *Norge* flight; now he was being blamed for their heroic countryman's disappearance and likely death.

As the train left the station, Nobile turned his attention to the stack of newspapers in his lap and began reading the slew of international articles that had been generated immediately following the *Italia* crash and during the subsequent rescues. Reading the papers was incredibly painful, and Nobile considered them as a whole to be "a shameful campaign of slander" launched against him. In addition to being blamed for Amundsen and the crew of five others now lost in the Latham 47 biplane, Nobile was being criticized for cowardice, for "running away" and abandoning his men by coming off the ice first; he was being held responsible for Finn Malmgren's death, with one editorial arguing that as commander, he should never have allowed Malmgren, Zappi, and Mariano to leave the Red Tent. There were even a few sensational articles suggesting that because of Nobile's decisions, Zappi and Mariano had resorted to cannibalizing Malmgren to save themselves. All of this, plus the deaths of the men who were carried away in the *Italia* envelope, was blamed on Umberto Nobile.

Nobile closed his eyes as the train chugged south. There was so much to process, so many thoughts and images swirling in his mind. When the survivors of the Red Tent had been transferred from the *Krassin* to the *Città di Milano,* there had been a brief and joyous reunion, but it had soon been tempered by harsh realities. Mariano was taken to a small "hospital room" on board, where his right foot was amputated just above the ankle. Nobile visited Mariano after the surgery, which he had endured with only local anesthetics. "He was lying in bed," Nobile wrote of seeing his first officer, "with the traces of indescribable sufferings clearly visible in his face. He told me about their march, and the recital of such dreadful adventures made my heart ache."

Zappi had also related his version of the ordeal, including the horrendous decision they were forced to make in leaving Finn Malmgren to die on the ice. Nobile had been moved to tears when Zappi described the final scene of them walking away, and although he was distraught at their decision, he found a way to comprehend it. "I understood," he wrote later, "how the two officers might have been placed in such a crucial position that they had been obliged to accept the separation."

As the train neared Sweden, Nobile cast back again to the men lost with the *Italia* envelope: Aldo Pontremoli, Ugo Lago, Ettore Arduino, Attilio Caratti, Callisto Ciocca, Renato Alessandrini. He conjured each man's face clearly, reflecting with bitterness that despite his desperate entreaties to keep looking for them, the *Krassin* had been instructed to discontinue its search efforts for these unfortunate men, the belief being that it would likely be in vain and unduly risk the lives of the rescuers in airplanes. Nobile was now resigned to never knowing what happened to them. Only "the polar ice," he realized, "holds the tragic secret of their fate."

So, it was with a heart burdened by guilt, self-doubt, shame, and bitterness that Nobile finished reading more than a month's worth of newspaper articles from across the world.

"The most ignominious things had been written about me and my companions," he observed. "I felt crushed under the avalanche of abuse hurled at me by mean spirits throughout the world . . . A great sadness took possession of me, a profound disgust with life and humanity."

At Vindeln, the first stop in Sweden, Nobile watched as a little "fair-haired, blue-eyed" girl approached the rail car. She was holding a big bouquet of flowers, and she reached out and handed them to him. He thanked her and became visibly emotional, choking up. After all the smears and slander heaped on him, he was moved by the sweet gesture of this "pure and gentle spirit." Her kindness in that moment revived his faith in the inherent goodness of humankind, and he vowed to never forget her name: Ebba Haggstrom.

Ever since the moment of the *Italia* disaster (and even well before, Nobile would come to discover), Prime Minister Benito Mussolini and his young and ambitious undersecretary Italo Balbo had been conspiring to distance themselves from Umberto Nobile and the international embarrassment that the crash and subsequent rescues had caused. According to most narratives being published, the story looked like this: Italy's flagship airship had been lost; General Nobile had abandoned his men; some of his men had deserted him and left one of their own to die, and then might well have cannibalized him. After all of that, the surviving Italians had ultimately been rescued not by their mother country but by Russians, Swedes, Norwegians, Finns, and Frenchmen. It was a public relations catastrophe, becoming among the largest media events of the entire 1920s.

Mussolini's Fascist government had been more than happy to take credit for Italy's role in the successful *Norge* flight two years earlier, but Nobile's *Italia* catastrophe, which garnered months of daily international criticism, was an event they wanted forgotten as soon as possible, even if that meant obliterating Nobile's reputation. So it was that Mussolini organized no formal reception for Nobile and his men, sent no representatives from Rome to greet them on their arrival. Despite these official snubs, as the train traveled through Italy, at every stop Nobile and his men saw platforms swarmed with enthusiastic onlookers welcoming them with cheers and showering them with flowers. Such spontaneous reception among the Italian populace occurred at every stop until they reached Rome on the evening of July 31, 1928.

Nobile could hardly believe what he encountered as he and the other courageous aeronauts disembarked. More than 200,000 people had gathered at and around the station area to welcome them home from their polar ordeal. It was, Nobile wrote of the scene, "a delirious crowd: a crowd which, hour by hour, had lived through the *Italia* expedition; had suffered the anguish of silence, the strain of suspense, the uncertainties of the rescue work; a crowd which had lived through all our tragedy, shared all our sufferings, felt all the infamies committed against us and our enterprise." Nobile was overwhelmed and humbled by the public outpouring, which assuaged, at least momentarily, his feelings of anguish, guilt, and sorrow for all that had been lost.

At the time of his arrival home, the *Italia* crash and rescue operations had claimed the lives of fourteen men: seven Italians, two Norwegians, a Swede, and four Frenchmen. Sadly, there would be more. In September, an Italian plane returning to Italy after vainly searching for the men lost in the *Italia* envelope crashed in France, killing three men. Nobile now had these men on his conscience as well. The crash of the *Italia* initiated the largest search and rescue mission in modern history, involving fourteen thousand men, twenty-three airplanes, sixteen ships, several dog sled teams and ski patrols, with eight countries participating.

But much of the criticism against Nobile focused on the disappearance of Roald Amundsen, who had courageously flown to Nobile's rescue. Amundsen's international fame ensured press interest in his vanishing, for which much of the world blamed Umberto Nobile. By late fall 1928, Nobile learned that Riiser-Larsen and American Louise Arner Boyd had finally given up their search for Roald Amundsen. Though they had covered ten thousand nautical miles in the *Hobby* over a ten-week period, they had found no trace of Roald Amundsen or the other five members of the crew of the Latham 47.

Eventually, a float from the airplane was discovered by a fishing vessel off northern Norway, as was a fuel tank. After these were determined to be from the Latham 47, it was concluded that the float plane had most likely crashed into the water or onto the ice, and that everyone aboard had perished. A few independent searches persisted for a time, but eventually it had to be admitted, even among the most hopeful Norwegians, that the White Eagle of Norway, the "Last Viking" Roald Amundsen, was dead.

Despite the acceptance and even adoration that Nobile received from the Italian public, on returning home he was shunned by Mussolini and the Fascist regime. Nobile was locked out of his office at the airship factory and told he was not to write articles, give public speeches or speak to the press, or even attend public events such as the theater. He and his wife, Carlotta, were put under surveillance to ensure their compliance. By the autumn of 1928, Umberto Nobile was definitively persona non grata, an airship designer without an airship to design, although he was technically still on the government payroll and serving as a general. To make matters even worse, eventually a seven-person committee (many of whom were hostile to Nobile) was formed to investigate the *Italia* disaster's causes.

The Commission of Inquiry was ultimately critical of Nobile, holding him, as commander, responsible for the crash. The results of the committee report were widely published, including with this scathing headline from a paper in Rome: "Nobile Blamed for *Italia* Disaster!" *The New York Times* was equally damning: "Nobile Is Censured in Official Report—Commission Blames Him for Loss of Dirigible *Italia* on Polar Flight."

The verdict of the committee stung Nobile and further destroyed his reputation, though many airship and polar experts came quickly to his defense, concluding that the findings were erroneous and unfair. Flying dirigibles in the Arctic was risky and dangerous, and the weather conditions, with high winds, fog, and icing had all likely contributed to the accident. Experienced Norwegian aviator and polar explorer Tryggve Gran was disgusted by Nobile's treatment in the report: "The judgment against Nobile . . . is unjust. In this case, Italy took recourse to finding a scapegoat. To say that Nobile has not conducted himself wholly honorably is preposterous . . ." Another Norwegian, author Sven Elvestad, studied the report and the statements by surviving members of the expedition. "When we compare all that has been written about the expedition and the catastrophe," Elvestad wrote, "including the accounts of other people besides Nobile, we gain a clear impression that the enterprise was well prepared and led."

Given the public disgrace, Nobile decided to resign from the military and become a private citizen. Italy was moving away from airship design anyway. They built only one more dirigible after the *Italia,* focusing their attentions instead on the airplane, which was innovating and vying for supremacy as the safest, most efficient and expedient means of air travel. Nobile, among the most skilled and experienced airship designers in the world, was recruited by the Soviet Union, and he went to Moscow and spent five years advising and consulting on their airship program. After the death of his wife, Carlotta, in 1934, Nobile traveled to the United States, where he taught courses and became head of the aeronautical engineering department of Lewis College of Science and Technology in Illinois for a few years.

Although he flourished in his work abroad, Nobile always deeply resented his forced exile and would spend much of the rest of his life trying to restore his reputation. It would not be until 1943—with the fall of Prime Minister Benito Mussolini and his Fascist regime—that Nobile would be able to fully explain his actions (now without government censorship) and refute the findings against him by the Italian Commission of Inquiry. He finally did so in a book titled *Now I Can Tell the Truth,* in which he argued that the inquiry against him was rigged. In 1945, a new commission exonerated him of blame for the *Italia* crash and restored his rank of major general of the air force. He was promoted to lieutenant general in 1946 and received back pay from the date of his resignation in 1928. The redemption meant a great deal to Nobile, who had suffered the emotional trauma of lives lost under his command. Now, at least the public knew that it was not his fault.[1]

Through all of the criticism and indignity, Umberto Nobile was comforted by the unwavering support he and his men received from Pope Pius XI, who called the historic Arctic flights of the *Italia* "worthy of universal admiration." He described Nobile's polar airship explorations as "feats which attain the highest beauty and sublimity that can be encountered in this life."

---

1 As recently as 2019, some scholars have attributed sleep deprivation and human fatigue as main causes of the *Italia* crash, pointing out that at the time of the disaster, Umberto Nobile had been awake for at least seventy-two hours. For his part, Nobile believed that icing of the hydrogen release valves precipitated their sudden plunge from the sky. It seems fair to conclude that both were to blame. See "Human Fatigue and the Crash of the Airship Italia," Gregg A. Bendrick, Scott A. Beckett & Elizabeth B. Klerman, *Polar Research,* December 2019, https://polarresearch.net/index.php/polar/article/view/3467/9984.

EPILOGUE

THE POLAR AIRSHIP AGE WAS AN AGE OF HEROES, AND THE MEN who flew airships in the Arctic between 1906 and 1928—aeronauts Walter Wellman, Roald Amundsen, Umberto Nobile—were the equivalent of the first astronauts. They were often treated as such, too, garnering international fame, ticker-tape parades, sometimes financial reward, occasional controversy, and polar exploration immortality.

Of the three visionaries, the pioneer Walter Wellman remains the least respected and the most misunderstood. One of Wellman's challenges was that of public perception: he was such a prolific and noted journalist, and so effective at promoting his adventures, that it was difficult for him to be taken seriously as an Arctic explorer, much less as an airship designer and navigator. The vaunted Norwegian explorer Fridtjof Nansen, for example, had this to say about Walter Wellman:

> A strange man who demonstrates how one may, with the help of the great art of advertising, keep the world's newspapers' attention year after year . . . In the course of several years (from 1906 to 1909), the newspapers were filled with constant dispatches about every step of this strange expedition's preparations, about all the glorious feats it was to accomplish and about all the most modern inventions that were to be used.

Wellman was indeed a brilliant self-promoter who exploited newspaper coverage and publicity to garner financial backing for his projects. Nansen and others criticized Wellman for his showmanship, suggesting he was more bluster than substance, but this was unfair. He made a half-dozen trips to the Arctic, cheating death more than once, at considerable personal cost to his finances, his career, and to his family. That he almost lost a leg in his attempts for the North Pole—and would walk for the rest of his life with a limp and a cane—suggests that he was as serious and committed as any other polar explorer. All of his contemporaries sought fame and glory, and so did Wellman. But we should not doubt his sincerity in attempting to advance scientific knowledge. He simply understood, before anyone else, that the old, tried and failed traditional dogsled method was past its time. Wellman was the first to comprehend that it would take motorized flight to unlock the pole.

As a polar explorer, Walter Wellman became associated with the word "failure," which is unfortunate and unjust. He attempted to reach the North Pole five times, and it is true that he failed every time—if failing means coming up short of the goal. It is also true that his polar flight attempts lacked sufficient preparation and were themselves effectively test flights. But I would counter that there was no other way to test fly an experimental airship such as Wellman's in the Arctic without actually doing it. In the early years of the twentieth century, Walter Wellman was conceiving, designing, and flying motorized hydrogen-filled aircraft hundreds of miles above the Arctic Circle, which was at the time as daunting and uncertain as the first manned space flights that would leave Earth's atmosphere a half century later.

Walter Wellman was criticized by his contemporaries (and has been by modern historians) for being primarily an entertainer who hyped up his expeditions through his wide media reach and then profited from them afterward by writing articles and books and speaking about the experiences. But nearly all Arctic explorers during Wellman's time—including Fridtjof Nansen, Robert Peary, and Dr. Frederick Cook, among many others—wrote books and conducted long lecture tours to profit from their experiences and hardships, as well as to advance the science of exploration. Wellman was merely part of the same tradition, and he should be recognized for adding the innovation of wireless communication to report his adventures in real time. Wellman sent the first wireless messages from the Arctic (to President Theodore Roosevelt!), and it was in fact this same technology by which a reporter aboard the *Norge* announced Amundsen, Ellsworth, and Nobile's arrival at the North Pole in 1926, and which would later—in 1928—allow Umberto Nobile to communicate his coordinates to the rescuers who would eventually save the *Italia* survivors.

Walter Wellman was a lot of things: a journalist, a political insider, an impresario, a celebrity, and an explorer. Mostly, Wellman was an innovator ahead of his time. What is fascinating and impressive is that he was simultaneously advancing the fields of journalism (through real-time reporting) and Arctic exploration (through technological innovation). Captivated by technology, he was also the first to attempt using motor-sledges—precursors to modern-day snowmobiles—in the high Arctic. Although the motor sledges did not work for Wellman, they illustrated his understanding that the days of the traditional dogsled explorers were ending. And of course, most significantly, he was the first to conceive of and attempt flying to the North Pole by airship. The courage this took, the vision and the panache, must be applauded. Although his own death-defying flights did not reach the top of the world, they showed explorers Roald Amundsen and Umberto Nobile that the airship was in fact the best, most efficient, and safest method to reach the summit of the globe. As early as 1893, Walter Wellman predicted in writing that "aerial navigation will solve the mystery of the North Pole and the frozen ocean." He was right.

It took until 1926—thirty-three years later—but Wellman's prediction proved prescient, and it provided him some retribution and validation that he was alive when it happened. He died not long afterward, in 1934, of liver cancer. But his vision had been realized during his lifetime by a Norwegian-led and American-financed expedition flying in an Italian-built airship. Not even the inimitable and prophetic Walter Wellman could have predicted such an international collaboration.

The unlikely pairing of Norwegian Roald Amundsen and Italian Umberto Nobile was a wedding of convenience, an arranged marriage whose honeymoon was the attainment of the North Pole by airship. Shortly thereafter, the union dissolved in a messy divorce, but the historical significance of the pairing continues to resonate through the exploring world today. When the airship *Norge* passed over and circled around the North Pole in the early morning hours of May 12, 1926, Roald Amundsen (as expedition leader) became the first to indisputably reach the North Pole, despite previous claims by Cook, Peary, and Byrd. Amundsen achieved in that moment the dual distinction of being first to both the South Pole and North Pole (along with his countryman Oscar Wisting), and the triumph vaulted Amundsen to the head of the pantheon of polar explorers. Sadly, during his lifetime, Amundsen would not receive the credit he deserved for being first to reach 90° N. It would take many decades to reveal that the claims by Cook, Peary, and Byrd were all flawed, either by outright falsification of records, omission of records, or highly exaggerated estimates of distances traveled. But none of the prior claims stood up to the intense scrutiny subjected to them by contemporary and later explorers, scholars, historians, and governing exploring organizations.

It is now widely accepted that Roald Amundsen was the first to the North Pole, the long-elusive holy grail of polar exploration. Having myself studied the various controversies and evidence, there is no doubt in my mind that Amundsen rightly deserves the laurels. It is also both tragic and ironic that Amundsen would live only two years beyond his attainment of the North Pole, and that in death, he would forever be linked to his public enemy Umberto Nobile and the disastrous flight of the *Italia*. And yet there is something fitting and romantic in Amundsen's "Hollywood" ending: He flew off into the Arctic mists to rescue his nemesis, never to be seen again. In certain respects, as callous as it sounds, it was the perfect heroic ending for Amundsen, cementing his legacy in storybook fashion. Amundsen is celebrated as a national hero in Norway, and in my estimation, he is the undisputed greatest polar explorer of all time.

As for Umberto Nobile, he has the dubious distinction, with his calamitous flight of the *Italia* in May of 1928, of essentially ending the era of polar exploration by airship. However, airship design, innovation, and experimentation—both for military and commercial purposes—continued into the next decade. The United States Navy built two impressive rigid airships—the USS *Akron* (ZRS-4) and the USS *Macon* (ZRS-5) in the early 1930s. Considered flying aircraft carriers, the helium-inflated airships carried small fighter biplanes which could be launched and recovered in flight, extending the range over which these behemoth (nearly eight hundred feet long) dirigibles could scout for and engage with enemy vessels. Although both of these aircraft served for a few years, each eventually crashed (*Akron* April 4, 1933—with loss of seventy-three men; *Macon* February 12, 1935—with loss of two men), precipitating the end of the rigid airship era in the United States.

From 1928–1937, the massive German airships *Graf Zeppelin* and *Hindenburg* ruled the skies. The *Graf Zeppelin*—named after airship pioneer Ferdinand von Zeppelin—made an astounding 590 flights totaling cumulatively over one million miles, including the first commercial transatlantic passenger flights and an historic round-the-world flight in 1929. In 1931, the *Graf Zeppelin* made a polar flight to rendezvous and exchange mail with the Russian icebreaker *Malygin* at Franz Josef Land. Aboard the *Malygin,* in an amazing confluence, was Umberto Nobile, who had gone along hoping to discover any trace of his six compatriots missing since the loss of the *Italia* in 1928. Unfortunately, no clues or evidence of the lost men were found, but the *Graf Zeppelin* did manage a successful water landing on the Arctic Ocean to meet for a short time with the Russian icebreaker, and *Graf Zeppelin* subsequently photographed and mapped Franz Josef Land extensively. When Nobile disembarked *Malygin* at the Russian mainland at Arkhangelsk, he reflected poetically on his brief return to the Arctic: "I was intoxicated by the extremely pure, freezing air. My eyes rested with deep joy on the immaculate purity of the snow that was broken up only by azure patches of ice . . . that feeling of absolute liberty: separation from all material things." This was how Umberto Nobile preferred to remember the polar regions—not lying on the ice for many weeks with a broken arm and leg, surrounded by airship carnage at the *Italia* crash site.

On May 6, 1937, *Hindenburg*—after a successful career of ten round-trip journeys to North America and seven to South America—approached the mooring mast at Lakehurst, New Jersey. On board the airship were sixty-one crew and thirty-six passengers. As *Hindenburg* made its final approach to the Naval Air Station, the airship burst into flames and dropped from the sky within thirty seconds, killing thirty-six people: thirteen passengers, twenty-two crew members, and one ground crewman. The catastrophe was captured on film and reported on the radio by Herb Morrison. The film and still footage, and Morrison's shaken and horrified reporting of the disaster (including the chilling commentary "Oh, the humanity!") became the iconic sounds and images of the airship age, destroying public confidence in the dirigible as a safe mode of passenger transportation.

Amazingly, Umberto Nobile would live long enough to see not only the crash of the *Hindenburg,* but the successful *Apollo 11* moon landing in 1969, the development of the first space shuttles by NASA in 1972, and the successful unmanned Mars landings of *Viking 1* and *2* in 1976. Nobile lived to be ninety-three years old, dying on July 30, 1978. Near the end of his life, Nobile—based on the tremendous aeronautical progress he had witnessed (and indeed, had been a part of)—wrote the following, which bears a note of nostalgia: "Today the sun has set on airships and . . . it is feared that it has set forever."

As it turns out, the sun seems to be rising on the airship once more.

POSTSCRIPT
Renaissance of the Airship

AT DAWN ON NOVEMBER 8, 2023, AN OYSTER-WHITE, HELIUM-inflated, titanic airship was wheeled out of a World War II—era hangar and onto the tarmac of NASA's Moffett Field in Mountain View, California, for flight-testing operations. The engineers, company representatives, and handful of journalists assembled—wearing fluorescent yellow vests and hardhats—were dwarfed beside the monstrous aircraft, which was longer than three Boeing 737s lined in a row nose to tail.

The aircraft being revealed fully for the first time was the 408-foot-long *Pathfinder 1,* the largest aircraft in the world since the *Graf Zeppelin* and *Hindenburg* in the 1930s. *Pathfinder 3* will be even larger, at a stupendous six hundred feet, the length of two football fields. But even more impressive than their size are their capabilities and potential in a variety of areas and applications. These and other airships like them, being built and tested in the United States, England, and France, are on the cusp of revolutionizing modern aeronautics.

A brave new airship era is upon us. We are going back to the future. Advanced technology; lighter, stronger, and greener materials; electric propulsion; and safer, non-combustible fuels (such as stable helium) have ushered in a new world of super high-tech, low-emissions airships that consume just 10 percent of the fuel used by large passenger airliners. Some of these modern airships are already here, and more are coming in the next year or two. With designs hearkening back to the earliest days of the great airships and the explorers who piloted them—but with the addition of twenty-first-century technology and materials like titanium, Kevlar, and carbon fiber—they will be used in humanitarian aid, logistics and construction (cargo and equipment transport), scientific exploration, and luxury passenger travel.

Sergey Brin, cofounder of Google, is a major player in the movement, driven to augment humanitarian aid and rescue operations while reducing aviation's carbon footprint to combat climate change. Brin's Lighter Than Air Research (LTA) is building a fleet of advanced airships now, with the *Pathfinder 1* approved for test flights by the FAA in late 2023. By early 2024 people will see the spectacular rigid dirigible flying over San Francisco Bay. The *Pathfinder*-series aircraft—with a dozen electric motors, four fin rudders, and wheeled landing gear—have remarkable capabilities, including top speeds of seventy-five miles per hour and the ability to hover in place and land in remote areas. Such performance makes them perfect for disaster relief: if runways, roads, or ports are inaccessible due to natural disasters like earthquakes or hurricanes, these airships can deliver what imperiled communities require; if cell phone towers become destroyed, damaged, or disabled, Brin's airships can hover above, providing cell service linkage. *Pathfinder 1* has payload capacity to carry four hundred eighteen-foot dome tents for emergency shelter, thirty-five hundred gallons of drinking water, and thirty thousand pounds of rice, as just a few examples.

LTA Research is hardly alone. U.S.-based Skunk Works, a research arm of Lockheed Martin, is designing airships to transport medical supplies to isolated locations across the globe. France's Flying Whales also plans humanitarian/medical ferrying by airship, which will be crucial given the ongoing wars and global humanitarian crises. Flying Whales' stated mission is to "unlock remote areas around the world," according to spokesman Romain Schalck. The company's *LCA60T* airship will be capable of lifting sixty tons of cargo, and Flying Whales' representatives are in discussions with the United Nation's World Food Program over planned humanitarian missions. The approach will be to fly over very difficult-to-access regions and unload food and medical supplies in cargo containers by winching them down out of the cargo bay with slings. The French company plans for a maiden flight in 2024, followed by two years of testing, then final certification. Since the certification for flying freight is less time-consuming than passenger certification, they expect to be up and flying by 2026, and hope to produce 150 airships over the next decade.

One of Flying Whales' divisions, called "Flying Care," aims to use the new, eco-friendly, state-of-the-art airships as energy-self-sufficient mobile hospitals, bringing not only medicine but medical professionals to isolated populations to provide pediatric, ophthalmologic, and dental care and even surgical interventions. According to Xavier Attrait, a noted traveling physician who has worked with various international emergency medical field hospitals, "The Flying Care mobile hospital is an ideal solution to provide high-level hospital services with state-of-the-art technology to the most remote communities that usually do not have access to healthcare. It's innovative, environmentally friendly, humanitarian, and cost-effective and will contribute to the WHO and UN goals of achieving universal health coverage for an additional 1 billion people worldwide by 2030."

Hybrid Air Vehicles (HAV), a UK-based leader in sustainable aircraft technologies, currently has preorders for twenty of its *Airlander 10* airships, with deliveries expected in 2026. The *Airlander 10*—originally built for the United States Army's Long Endurance Multi-Intelligence Vehicle Program (LEMV) and first flown in the United States in 2012—has been upgraded and repurposed. It now has a ten-ton payload, a four-thousand-mile range, a maximum altitude of twenty thousand feet, and can remain airborne for five days. HAV plans to use the *Airlander 10* airships for communications, logistics, experiential travel, and to connect communities that are currently underserved by traditional mass transit, without the need for developed infrastructure like train tracks or landing runways. *Airlander* craft can take off and land on unprepared surfaces, allowing access to previously inaccessible places where there are no roads or airfields.

A few companies—like OceanSky Cruises—are planning private passenger travel within the next few years. Fleets of luxury airships (sometimes called "sky yachts") will have master suites, offices, lounges, and spas, and will be used for island hopping, safaris, and cruises over Svalbard, Norway, the very area explored by Wellman, Amundsen, and Nobile. OceanSky Cruises has ambitious plans for airship excursions in Africa and the Arctic. An airline company rather than an airship manufacturer, they view airship travel as more analogous to train travel in that the speeds will be slower and the ride will focus on space, comfort, cuisine, and spectacular views, with large windows and viewing galleries. Their proposed Southern Africa itinerary involves a round-trip journey from the Skeleton Coast in Namibia to Victoria Falls in Zimbabwe, touching down in inaccessible regions and exploring in Namibia, Botswana, and Zimbabwe.

But most intriguing—at least to me—is OceanSky Cruises' planned North Pole Expedition, which will take off and return from Longyearbyen, Svalbard's largest city. The two-day round-trip journey will include landing at the North Pole and walking around on the polar surface—something that not even Roald Amundsen was able to accomplish in 1926. According to Oceansky's marketing director, Gonzalo Gimeno, "For us, the North Pole is ground zero for climate change, and landing an airship there following Nobile and Amundsen's footsteps, for the first time in history, will send a very strong signal for the need for a new era in sustainable aviation and travel." The cost of the North Pole trip is $200,000 for forty-eight hours, which is a little steep, but I intend to finagle my way on a flight somehow, and to write about it. After all, what would Walter Wellman, Roald Amundsen, and Umberto Nobile do?

ACKNOWLEDGMENTS

Numerous people deserve mention for their invaluable assistance, support, and love during the research and writing of *Realm of Ice and Sky*. Every book I write is a new experience and journey, but one constant is that I could not do what I do and live the writing life I live without a lot of help.

Thanks to Scott Waxman, my literary agent, collaborator, and friend for over two decades. Scott's unwavering belief in my writing and ideas through the years has nurtured me and kept me going. His knowledge of the publishing industry is unrivaled, and I have greatly benefitted from his tutelage and sagacity. I especially enjoyed my time at the Larchmont Yacht Club in December of 2022 for the release of *Empire of Ice and Stone* and meeting his wonderful family.

At Waxman Literary Agency I've been brilliantly shepherded by agent Susan Canavan, who is as skilled, experienced, and savvy as anyone in the publishing business. I appreciate Susan's sense of humor, openness, and "straight shooter" approach and her dogged negotiation on my behalf. Mostly, she's just awesome.

I am fortunate to have a fantastic team at St. Martin's Press, spear-headed by my editor, Marc Resnick. Marc has been incredible to work with on three books so far, and I hope to collaborate with him on many more, at least until AI takes over and begins writing them for me. (Wait, that's not funny!) I value Marc's intuition, his deep understanding of strong storytelling, and of course, his keen eye and ear during the editing process. Thanks for your patience and understanding, Marc, of what writers need to produce their best.

At St. Martin's Press, the entire squad helps me at every phase of the process, and my books are true collaborations. Hats off to Lily Cronig, assistant editor; Sara Beth Haring, senior marketing manager; and Rebecca Lang, associate director of publicity. You are a dream team and you help make beautiful books of which I am very proud.

Many individuals and institutions were instrumental during my research. Washington State University, where I have taught and conducted research for thirty-five years (yikes), continues to support my work. Thanks to Donna Potts (Chair, English) and the College of Arts and Sciences for providing me with a 2023 Summer Research Fellowship, which contributed significantly to my research travel to Norway (Oslo and Svalbard) in June 2023, allowing me to visit a number of the places described in *Realm of Ice and Sky*.

In Norway, numerous shout-outs are owed. I am especially indebted to Anders Bache, curator at the Roald Amundsen House, Uranienborg. Anders gave my wife, Camie, and me an incredible private tour of the house on May 26, 2023. It was an honor to spend so much time there and to benefit from Anders's passion and deep understanding of Amundsen as not just an explorer but a man, one with flaws like the rest of us. Beyond the unforgettable tour, Anders has been an inexhaustible resource about Amundsen, airships, and all aspects of polar exploration.

In Tromsø, I received a wonderful two-hour tour of the Polar Museum (Polarmuseet) from museum advisor Vilde Ørsje Utby. Vilde is enthusiastic and erudite, and clearly loves her job. Thanks for bringing tears to my eyes with your beautiful telling of the Nils Strindberg and Anna Charlier tragic love story.

In Svalbard, I had the great fortune to receive a residency at the Spitsbergen Artists Center in Nybyen. What a fantastic opportunity to read and write in the historic mining outpost, with close proximity to the mountains and fjords of Longyearbyen. Thanks so much to the center's director, Elizabeth Bourne, who accepted my research proposal and arranged for my presentation at the Svalbard Museum. My time at the Spitsbergen Artists Center was made especially memorable by my fellow residents

Carolina Redondo Fernandez and Nina Schipoff. I will always value our meals and long talks about writing, art, film, and life! Only wish I had had more time with Carolina but hope to see her sometime in Madrid. Nina, what could possibly top my last night in Svalbard and the magnificent and moving Patti Smith concert? See you sometime in Switzerland.

My first week in Spitsbergen was spent at Longyearbyen Camping, which made for a very authentic Arctic experience as I slept in a "Scott" tent, one originally designed by Robert Falcon Scott. I will never forget the diving and attacking Arctic terns and the constant squabbling of barnacle geese. At Longyearbyen Camping I did my first (and so far only) Naked Polar Swim. Thanks so much to owner Michelle Van Dijk and especially to Torlief, the camp manager, with whom I had many illuminating conversations.

I wish to extend heartfelt thanks and love to my dear friend Ingerid Bøyum Aase. I first met Ingerid in Greenland in 2003 while I was covering a multi-day endurance competition in which she was competing. Ingerid eventually visited my family in Idaho, and we have remained friends for more than twenty years. Ingerid gave me a copy of Fridtjof Nansen's *First Crossing of Greenland,* and it was this book that got me enthralled with polar exploration literature. Ingerid kindly hosted my wife, Camie, and me at her home in Son, Norway, and at her lovely summer cabin. We had such an exhilarating visit and enjoyed the sea kayaking, walks, and special time with your family. We appreciate the meals of reindeer tenderloins, lingonberries, and Scandinavian shrimp sandwiches. Of course, our excursions in Svalbard were highlights—thanks for always being up for an adventure.

John Larkin, my dear friend and intrepid (and lightning-fast) first reader once again provided excellent initial edits, suggestions, and questions. His humorous remarks help lighten the editing process. Thanks Juan. Your porch is where it all begins. And much love and appreciation, too, to Melissa Rockwood, who is as constant a friend as one could have. Love you, Missy.

To my fellow Free Range Writers Kim Barnes, Collin Hughes, Jane Varley, and Lisa Norris: What a long, strange trip, eh? Glad we reconfirmed the location of the ammo can and hope not to unearth it very soon.

Smooches to my siblings—Lisa, Lance, and Lex. Glad we've managed to remain tight.

Finally, I am indebted most of all to my wondrous family. My wife, Camie, has been with me through every book (nine now?), and her patience and support never wavers (okay, it *almost* never wavers). She helps me remain strong and focused, especially when I'm on a deadline. Thanks to my beautiful daughter, Logan (and her hubby, Chris), for gifting me with gorgeous grandchildren Luke and Palmer (brand-new baby Polly Drew!) and helping me to remember what is really most important in life. And deep respect and love to my son, Hunter, who has become a prolific, contemplative reader, writer, and impressive critical thinker. I value our ongoing conversations about literature, politics, and life.

Thanks to everyone. Now, on to the next book . . .

AUTHOR'S NOTE ON THE TEXT AND SOURCES

The lives, times, and aerial expeditions of Roald Amundsen and Umberto Nobile have been extensively chronicled; Walter Wellman's story considerably less so. I must admit that prior to beginning research on this book, I had never heard Wellman's name. The more research I conducted, the more I began to see how Wellman's efforts between 1906–1909 set the stage for the epic drama that would ultimately play out when Amundsen and Nobile determined that the airship was indeed the best method to achieve the North Pole. The first third of the twentieth century was a remarkable and transitional time in polar exploration history—essentially the end of the traditional sailing ship and dogsled approach giving way to aerial exploration—and I became mesmerized by all three of these men (and their intrepid crews) who had the vision and courage to sail skyward in their quests for scientific discovery, fame, and fortune.

To the extent that it is possible, I endeavor to travel to the places I write about in order to better understand and be able to describe the landscapes (and icescapes) that the historical figures I'm writing about encountered. Such research travel is essential to me, and it is without question the most enjoyable part of the research process. In June 2023, I traveled to Svalbard for this reason, and I was awed and humbled by the spectacular archipelago. I was able to witness polar bears in the wild at close range (from a ship!), watching their astonishing grace and speed as they hunted along the shoreline of the Billefjorden near Pyramiden. Seeing their immense power as they bashed through the thick ice foot trying to get to seals gave me a sense of what polar explorers stranded on the ice had to contend with.

Most significant and illuminating was my trip to visit Ny-Ålesund (Kings Bay), the place from which Amundsen, Ellsworth, Nobile (and Byrd and Bennett) embarked on their aerial voyages. The one-hundred-nautical-mile boat ride from Longyearbyen to Ny-Ålesund (five hours each way) was rough and spine-jarring in heavy seas, but well worth it to visit the northernmost permanently inhabited town in the world at 78° 56′ N. I was able to stand at the mooring mast to which the *Italia* once tethered, to see where the great airship hangar used to be located, and to visit the Amundsen Villa where Roald Amundsen stayed during his time there. Exploring Kings Bay and environs on foot, I took in the sprawling sweep of the mountains, seas, and glaciers—it was all magnificent and illuminating and provided me with essential perspective of this wondrous, far-flung place.

In researching and writing *Realm of Ice and Sky* I have relied on hundreds of historical newspaper articles, firsthand books and articles by the participants, secondary source books and scholarly papers, personal interviews, documentary films, and visits to domestic and international museums and archives. The literature of polar exploration is voluminous, the niche field of airship polar exploration less so, but still rich and fascinating. For readers wishing to discover more and delve deeper into the subject, relevant selected works are listed in the Document Collections and Selected Bibliography that follows. These works have been quoted directly, cited, or consulted as valuable references in the writing of *Realm of Ice and Sky*.

Walter Wellman (November 3, 1858–January 31, 1934) was a household name during the peak of his journalism career, which dovetailed with his exploration career since he was a pioneer in writing about and covering as news his own exploits, yet he remains virtually unheard of today. The work of two scholars in particular has greatly contributed to my understanding of Walter Wellman as a man, a journalist, and an explorer. P. J. Capelotti, Professor of Anthropology at Pennsylvania State University, is a dedicated and prolific scholar who has written the most in-depth and salient work on Walter Wellman that currently exists. I have relied on his erudite, incisive, and expansive books and articles, and a few deserve special mention for forming and solidifying my understanding of Wellman's life, explorations, airship design, Arctic wireless innovations, and contributions to polar history. Capelotti's book *The Greatest Show in the Arctic: The American Exploration of Franz Josef Land, 1898–1905* (University of Oklahoma Press, 2016) provided me with a rich and nuanced understanding of Walter Wellman's early life as well as his initial polar explorations by traditional method. The six-hundred-plus-page work illustrates scholarship of the highest order and an intriguing study of not only Wellman's adventures and misadventures in Franz Josef Land but also two other American expeditions that are relatively unknown—the Baldwin-Ziegler (1901–02) and the Fiala-Ziegler (1903–05).

Two other significant works by P. J. Capelotti were instrumental to the Walter Wellman section (Part One) of *Realm of Ice and Sky*. The first is the monograph study *The Wellman Polar Airship Expeditions at Virgohamna, Danskøya, Svalbard: A Study in Aerospace Archaeology* (Oslo: Norsk Polar-Institutt, 1997). This comprehensive and provocative study—the result of Capelotti's extensive academic and field research at the site of Wellman's base camp at Virgohamna (Danes Island, Svalbard)—provides vital historical context for Wellman's airship attempts as well as detailed schematics of the airship *America* versions that Wellman built and used.

Last (in addition to Capelotti's numerous articles and proceedings papers), Capelotti's *By Airship to the North Pole: An Archaeology of Human Exploration* (New Brunswick, N.J. Rutgers University Press, 1999) proved invaluable to me. Particularly useful were his Select Chronology of Northern Expeditions and Events (from 1194–1978) and his penetrating analysis and conclusions about Wellman's place in polar airship history and Wellman's conjoining of technological innovation and the mass media in Arctic exploration. I relied on P. J. Capelotti's works greatly in writing Part One: The Pioneer—The Remarkable Walter Wellman—and that section of *Realm of Ice and Sky* would be much diminished without Capelotti's decades of field work and scholarship.

The other important Wellman historian (of which there are few) whose work I relied on heavily is David L. Bristow, editor of *Nebraska History Magazine*. Bristow's book *Flight to the Top of the World: The Adventures of Walter Wellman* (University of Nebraska Press. Lincoln, Neb.: 2018) is, to my knowledge, the only full-length biography on Wellman. Bristow's book is comprehensive and detailed, providing nuance and insight into Wellman's journalistic career and family life that was essential to my understanding of Wellman as a person. Particularly salient were the extensive sections on Wellman's life as a Washington political insider and his impact on American journalism. My treatment of Wellman would be much diminished without Bristow's excellent book as essential source material. Additionally, I've based Wellman's flight path on September 2, 1907 (depicted in Jeffrey L. Ward's map at front of this book) on Bristow's own map found on page 159 of his book. Like Bristow, I relied primarily on Felix Riesenberg's description and account in his 1937 book *Living Again: An Autobiography*.

Roald Amundsen's life and explorations have been extensively documented (and in Norway, lionized), but one relatively recent book deserves mention here for its thoroughness, evenhandedness, and storytelling excellence. Stephen R. Bown's *The Last Viking: The Life of Roald Amundsen* (Da Capo Press. Boston: 2012) is a superb work of narrative history, a cradle-to-grave account of Amundsen that is laudatory of the man where warranted but pulls no punches where Amundsen deserves harsh criticism. Bown, a Canadian, possesses a wealth of knowledge about the history of Arctic exploration, and there is a surefootedness to his engaging storytelling. *The Last Viking*—while at times reading like a thrilling adventure novel— helped me to understand not only what Amundsen did but the underlying reasons and rationale for many of the choices he made—some of which hurt his public persona and destroyed relationships. Mr. Bown, who has written a dozen books, many of which concern exploration and discovery, has penned the definitive Amundsen biography—one I highly recommend and one that I leaned on for quotes and as a fundamental resource.

As much as possible, I drew directly on the works written by the participants of the events recounted and described in *Realm of Ice and*

Sky. Noteworthy are the following first-hand accounts: *The Aerial Age* by Walter Wellman (A.R. Keller. New York: 1911); *My Life as An Explorer* (Doubleday. Garden City, N.Y.: 1927) by Roald Amundsen; *First Crossing of the Polar Sea* (Doubleday, Duran and Company. Garden City, N.Y.: 1928) by Roald Amundsen and Lincoln Ellsworth; *With the* Italia *to the North Pole* (George Allen and Unwin LTD. London: 1930), and *My Polar Flights* (G. P. Putnam's Sons. New York: 1961) by Umberto Nobile; *Beyond Horizons* (Doubleday. London: 1938) by Lincoln Ellsworth; *Living Again: An Autobiography* (Doubleday. London: 1937) by Felix Riesenberg; *Norwegian Bravado: Diaries From Franz Josef Land and Danes Island* (Vagemot Miniforlag, Oslo: Norway, 2004) by Paul Bjørvig; *The Arctic Rescue: How Nobile Was Saved* (Viking Press. New York: 1929) by Einar Lundborg; and *The Tragedy of the Italia: With the Rescuers to the Red Tent* (London, Ernest Benn LTD, 1928) by Davide Giudici.

The selected bibliography that follows offers more detail with regards to these works and represents much of the material I consulted and sourced while writing *Realm of Ice and Sky.*

DOCUMENT COLLECTIONS AND SELECTED BIBLIOGRAPHY

Document Collections, Museums and Archives

Explorers Club, New York

Fram Museum, Oslo, Norway

Grenna Museum, Andrée Expedition (Gränna, Sweden): https://www.grennamuseum.se/en/the-andree-expedition/

Italian Air Force Museum, Vigna di Valle, Italy

Library of Congress, Washington, D.C.

Maritime Museum of British Columbia, Victoria, British Columbia

National Maritime Museum, Amsterdam, Netherlands

New York Public Library

North Pole Expedition Museum, Longyearbyen, Svalbard, Norway

Norsk PolarInstitutt (Norwegian Polar Institute): https://www.npolar.no/en/

Ny-Ålesund Museum, Svalbard, Norway

Polar Museum, Tromsø, Norway

Roald Amundsen House, Uranienborg, Norway: https://mia.no/roaldamundsen/en

Smithsonian Institution—National Air and Space Museum Archives

Svalbard Museum, Longyearbyen, Svalbard, Norway

Walter Wellman Collection, DSO-2577–3487. Library of Congress, Washington, D.C.

Walter and Arthur Wellman Collection, Smithsonian National Air and Space Museum: https://transcription.si.edu/project/17980

Historical Newspaper Resources

America's Historical Newspapers, Readex: A Division of NewsBank
California Digital Newspaper Collection
Chronicling America: Library of Congress
The New York World (Digital Delivery System)
TimesMachine: *The New York Times*.
Washington State Library and Archives: ILLIad/Interlibrary Loans

Books and Official Reports

Amundsen, Roald. *My Life As an Explorer*. Garden City, New York: Doubleday, 1927.

Amundsen, Roald and Lincoln Ellsworth. *First Crossing of the Polar Sea*. Garden City, New York: Doubleday, 1928.

——*Our Polar Flight* (1925).

Andrée, S.A. *Andrée's Story: The Complete Record of his Polar Flight, 1897*. New York: Viking Press, 1930.

Arneson, Odd. *The Polar Adventure: The "Italia" Tragedy Seen at Close Quarters*. London: Victor Gollancz, 1929.

Balchen, Bernt. *Come North With Me: An Autobiography*. New York: Dutton, 1958.

Baldwin, Evelyn Briggs. *The Search for the North Pole*: 1896 https://archive.org/details/cihm_57137.

Bart, Sheldon. *Race to the Top of the World: Richard Byrd and the First Flight to the North Pole*. Washington, D.C.: Regnery, 2013.

Beaubois, H. *Airships: Yesterday, Today and Tomorrow*. New York: Two Continents Publishing Company, 1973.

Berton, Pierre. *The Arctic Grail*. New York: Viking, 1988.

Bjorvig, Paul. *Norwegian Bravado: Diaries from Franz Josef Land and Danes Island*. Trans. by Sue Nichols. Skein, Norway Vagemot Miniforlag, 2004.

Bomann-Larsen, Tor. *Roald Amundsen*. Gloucestershire, UK: The History Press, 2011.

Bown, Stephen. *The Last Viking: The Life of Roald Amundsen*. Boston, MA: Da Capo Press, 2012.

Botting, D. *Dr. Eckener's Dream Machine: The Great Zeppelin and the Dawn of Air Travel*. New York: Henry Holt and Company, 2001.

Bristow, David. *Flight to the Top of the World: The Adventures of Walter Wellman*. Lincoln, NE: University of Nebraska Press, 2018.

Bryce, Robert M. *Cook & Peary: The Polar Controversy Resolved*. Mechanicsburg, PA: Stackpole Books, 1997.

Cameron, Garth. *From Pole to Pole: Roald Amundsen's Journey in Flight*. New York: Skyhorse, 2013. *Umberto Nobile and the Arctic Search for the Airship* Italia. Croydon, UK: Fonthill 2017.

Capelotti, P.J. *The Greatest Show in the Arctic: The American Exploration of Franz Josef Land, 1898–1905*. Norman, OK: University of Oklahoma Press, 2016.

——*By Airship to the North Pole: An Archeology of Human Exploration*. New Brunswick, N.J.: Rutgers University Press, 1999.

——*The Wellman Polar Expeditions at Virgohamna, Danskøya, Svalbard; a Study in Aerospace Archeology*. Meddeslelser no 145. Oslo: Norwegian Polar Institute, 1997.

Cross, Wilbur. *Disaster at the Pole: The Crash of the Airship* Italia—*A Harrowing True Tale of Arctic Endurance and Survival*. Guilford, CT: The Lyons Press, 2002.

Deighton, Len and Arnold Schwartzman. *Airshipwreck*. New York: Holt, Rinehart, & Winston, 1979.

Dithmer, Elisabeth. *The Truth About Nobile* (translated from the Danish by Elisabeth Dithmer and Frank Fleetwood). London: Williams & Northgate, 1933.

Drivenes, Einar-Arne and Harald Dag Jolle. *Into the Ice: The History of Norway and the Polar Regions*. Oslo: Gyldendal, 2006.

Ellis, Richard. *On Thin Ice*. New York: Alfred A. Knopf, 2009.

Fiala, Anthony. *Fighting the Polar Ice*. New York: Doubleday, 1907.

Fleming, Fergus. *Ninety Degrees North: The Quest for the North Pole*. New York: Grove Press, 2001.

Francis, Daniel. *Discovery of the North*. Edmonton, Alberta, Canada: Hurtig, 1986.

Geoghegan, John. *When Giants Ruled the Sky: The Brief and Tragic Demise of the American Rigid Airship*. Gloucestershire, UK: The History Press, 2022.

Gertner, Jon. *The Ice at the End of the World: An Epic Journey into Greenland's Buried Past and Our Perilous Future*. New York: Random House, 2019.

Gosnell, Mariana. *Ice: The Nature, History, and the Uses of an Astonishing Substance*. New York: Alfred A. Knopf, 2005.

Giudici, Davide. *The Tragedy of the* Italia: *With the Rescuers to the Red Tent*. London: Ernst Benn LTD, 1928.

Glines, C.V. *Polar Aviation*. Eau Claire, Wis.: E.M. Hale and Company, 1967.

Goldberg, Fred. *Drama in the Arctic: The Search for Nobile and Amundsen: A Diary and Postal History*. Lidingo, Sweden: The Fram Museum, 2003.

Grierson, John. *Challenge to the Poles*. London: G.T. Foulis & Co., LTD, 1964.

Gwynne, S.C. *His Majesty's Airship: The Life and Tragic Death of the World's Largest Flying Machine*. New York: Scribner, 2023.

Hartman, Darrell. *Battle of Ink and Ice: A Sensational Story of News Barons, North Pole Explorers, and the Making of Modern Media*. New York: Viking, 2023.

Herbert, Wally. *The Noose of Laurels: Robert E. Peary and the Race to the North Pole*. New York: Atheneum, 1989.

Hogg, Gary. *Airships Over the Pole: The Story of the* Italia. London: Abelard-Shuman, 1969.

Holland, Clive, ed. *Farthest North: A History of North Polar Exploration in Eye-Witness Accounts*. New York: Carroll & Graf, 1994.

Holland, Eva. *Mussolini's Arctic Airship* (2017 Kindle Single).

Hovdenak, Gunnar. *Roald Amundsens Siste Ferd*. Oslo: Gyldendal, 1934.

Huntford, Roland. *The Last Place on Earth: Scott and Amundsen's Race to the South Pole*. New York: The Modern Library, 1999.

Jackson, Frederick G. *A Thousand Days in the Arctic*. London: Harper & Brothers, 1899.

Kafarowski, Joanna. *The Polar Adventures of a Rich American Dame: The Life of Louise Arner Boyd*. Toronto, Canada: Dundurn, 2017.

Kaplan, Susan A. and Genevieve M. LeMoine. *Peary's Arctic Quest: Untold Stories from Robert E. Peary's North Pole Expeditions*. Camden, Maine: Down East Books, 2019.

Levy, Buddy. *Labyrinth of Ice: The Triumphant and Tragic Greely Polar Expedition*. New York: St. Martin's Press, 2019.

——*Empire of Ice and Stone: The Disastrous and Heroic Voyage of the* Karluk. New York: St. Martin's Press, 2022.

Leonard, Max. *A Cold Spell: A Human History of Ice*. New York: Bloomsbury, 2024.

Loomis, Chauncy. *Weird and Tragic Shores*. New York: Knopf, 1971.

Lopez, Barry. *Arctic Dreams: Imagination and Desire in a Northern Landscape*. New York: Scribner, 1986.

Lundborg, Einar. *The Arctic Rescue*. New York: Viking, 1929.

Lundstrom, Sven. *Andrée's Polarexpedition*. Gränna: Wiken, 1988.

Mabley, Edward. *The Motor Balloon* "America." Brattleboro, VT: The Stephen Greene Press, 1969.

Maynard, Jeff. *Wings of Ice: The Riddle of the Polar Air Race*. London: Lume Books, 2020.

McCorristine, Shane. *The Spectral Arctic: A History of Dreams and Ghosts in Arctic Exploration*. London: UCL Press, 2018.

McCullough, David. *The Wright Brothers*. New York: Simon & Schuster, 2015.

McGlynn, Frank. *Stanley: The Making of an African Explorer*. London: Constable, 1989.

McKee, Alexander. *Ice Crash: Disaster in the Arctic, 1928*. New York: St. Martin's Press, 1979.

Meyer, H.C. *Airshipmen, Businessmen, and Politics, 1890–1940.* Washington, D.C.: Smithsonian Institution Press, 1991.

Mirsky, Jeannette. *To the Arctic!: The Story of Northern Exploration from the Earliest Times.* Chicago: University of Chicago Press, 1970.

Mittelholzer, Walter. *By Airplane Towards the North Pole.* Boston: Houghton Mifflin, 1925.

Montagu, Lord and B. Baden-Powell. *A Short History of Balloons and Flying Machines.* North Haven, CT: Read Books Ltd, 2013.

Montague, Richard. *Oceans, Poles, and Airmen: The First Flights Over Wide Waters and Desolate Ice.* New York: Random House, 1971.

Moss, Sarah. *The Frozen Ship: The Histories and Tales of Polar Exploration.* New York: Bluebridge, 2006.

Nansen, Fridtjof. *In Northern Mists.* 2 Vols. London: William Heinemann, 1911.

——*Farthest North: The Epic Adventure of a Visionary Explorer.* New York: Skyhorse Publishing, Inc., 2008.

Neatby, Leslie H. *Conquest of the Last Frontier.* Athens: Ohio University Press, 1911.

Nobile, Umberto. *My Five Years with Soviet Airships.* Akron, Ohio: Lighter-than-Air Society, 1987.

——*With the* Italia *to the North Pole.* London: George Allen & Unwin, LTD, 1930.

——*My Polar Flights.* New York: G. P. Putnam's Sons, 1961.

Nuttal, Mark, ed. *Encyclopedia of the Arctic.* New York: Routledge, 2005.

Parijanine, Maurice. *The* Krassin. New York: The Macaulay Company, 1929.

Peary, Robert. *The North Pole.* New York: Frederick A. Stokes, 1910.

Piesling, Mark. *N-4 Down: The Hunt for the Arctic Airship* Italia. New York: Custom House, 2021.

Pool, Beekman H., *Polar Extremes: The World of Lincoln Ellsworth.* Fairbanks: University of Alaska Press, 2002.

Pitzer, Andrea. *Icebound: Shipwrecked at the Edge of the World.* New York: Scribner, 2021.

Putnam, George Palmer. *Andrée: The Record of a Tragic Adventure*. New York: Brewer & Warren, 1930.

Rawlins, Dennis. *Peary at the North Pole: Fact or Fiction?* Washington, D.C.: Robert B. Luce, 1973.

Riffenburgh, Beau. *The Myth of the Explorer: The Press, Sensationalism, and Geographical Discovery*. New York: Oxford University Press, 1994.

Riesenberg, Felix. *Living Again: An Autobiography*. New York: Doubleday, 1937.

Roberts, David. *Alone on the Ice: The Greatest Survival Story in the History of Exploration*. New York: W. W. Norton, 2013.

Robinson, Douglas H. *Giants in the Sky: A History of the Rigid Airship*. Seattle: University of Washington Press, 1973.

Rose, Alexander. *Empires of the Skies: Zeppelins, Airplanes, and Two Men's Duel to Rule the World*. New York: Random House, 2020.

Sancton, Julian. *Madhouse at the End of the Earth: The Belgica's Journey into the Dark Antarctic Night*. New York: Crown, 2021.

Santos-Dumont, Alberto. *My Airships*. New York: Century, 1904.

Simmons, George. *Target: Arctic. Men in the Skies at the Top of the World*. Philadelphia, PA: Chilton, 1965.

Simmonds, Peter Lund. *Sir John Franklin and the Arctic Regions*. Auburn and Buffalo, New York: Miller, Orton and Mulligan, 1844.

Sollinger, Günther. *S.A. Andrée: The Beginning of Polar Aviation, 1895–1897*. Moscow: Russian Academy of Sciences, 2005.

Stark, Peter. *Ring of Ice: True Tales of Adventure, Exploration, and Arctic Life*. New York: Lyons Press, 2000.

Streever, Bill. *Cold: Adventures in the World's Frozen Places*. New York: Little, Brown, 2009.

Sundman, Per Olaf. *The Flight of the Eagle*. New York: Pantheon Books, 1970.

Tandberg, Rolf. *The* Italia *Disaster: Fact and Fantasy* Privately printed: Oslo, 1977.

Tessendorf, K.C. *Over the Edge: Flying with the Arctic Heroes*. New York: Atheneum, 1998.

Thomas, David N. *Frozen Oceans: The Floating World of Pack Ice*. London: Firefly Books, 2004.

Wallance, Gregory J. *Into Siberia. George Kennan's Epic Journey Through the Brutal, Frozen Heart of Russia*. New York: St. Martin's Press, 2023.

Wellman, Walter. *The Aerial Age*. New York: A. R. Keller, 1911.

Wilkins, George. *Flying the Arctic*. New York: Grosset & Dunlap, 1928.

Wilkinson, Alec. *The Ice Balloon*. New York: Knopf, 2012.

Williams, Glyn. *Arctic Labyrinth: The Quest for the Northwest Passage*. Berkeley: University of California Press, 2009.

ARTICLES

Aas, Steiner. "New Perspectives on the *Italia* tragedy and Umberto Nobile." *Polar Research* 24, no. 1–2 (2005), 5–15.

Alger, George W. "The Literature of Exposure." *Atlantic Monthly* (August 1905): 210–13.

Arthur, Charles W. "Walter Wellman: Who Plans to Reach the North Pole in an Airship." *The World To-Day* 10, no. 3 (March 1906): 316–319.

Bendrick, Gregg A., Scott A. Beckett and Elizabeth Kierman. "Human Fatigue and the crash of the airship *Italia*." *Polar Research* 35 (July). https://doi.org/10.3402/polar.v35.27105.

Capelotti, P. J., Herman Van Dyk and Jean-Claude Caillez. "Strange Interlude at Virgohamna, Danskoya, Svalbard, 1906: The Merkelig Man, the Engineer and the Spy." *Polar Research* 6, no. 1 (2007): 64–75.

Cook, Frederick A. "The Conquest of Mount McKinley." *Harper's Magazine Monthly*, May 1907: 821–37.

Diesen, Jan Anders, and Neil Fulton. "Wellman's 1906 Polar Expedition: The Subject of Numerous Newspaper Stories and One Obscure Film." *Polar Research* 26 (2007): 76–85.

Egan, Maurice Francis. "Dr. Cook in Copenhagen: A Foot-Note to History." *Century Magazine*, September 1910: 759–63.

Herbert, Wally. "Commander Robert E. Peary: Did He Reach the Pole?" *National Geographic*, September, 1988: 346–413.

Hunt, William R. "Confrontation in Montana: Frederick A. Cook and the Mount McKinley Controversy." *Montana: The Magazine of Western History,* 34 no. 1 (Winter 1984): 46–56.

Kafarowski, J. "Searching for Amundsen: Louise Arner Boyd aboard the *Hobby*." *Sea History,* Winter Issue, no. 177 (2021): 12–17.

Nansen, Fridtjof. "How Can the North Polar Region Be Crossed?" *Geographical Journal* 1, no. 1 (January 1893): 1–22.

Nobile, Umberto. "The Dirigible and Polar Exploration." *The American Geographical Society* 7, no. 7, 1928.

Russell, Isaac. "Melvin Vaniman—Aerial Adventurer." *Collier's* 48, no. 9 (November 25, 1911): 17.

Stevens, George E. "Walter Wellman: Journalist, Explorer, 'Astronaut.'" [Lake County Ohio] *Historical Society Quarterly* 11, no. 3 (1969).

Vaniman, Melvin. "The New Dirigible—Its Great Possibilities." *Aero Club of America Bulletin* 1, no. 4 (May 1912): 9–11.

——"Revolutionizing Air Travel." *Aircraft,* May 1912, 76–78.

——"The Voyage of the *America*." *Aeronautics* 7 (December 1910): 199.

Wellman, Walter. "Amundsen's Advice to Wellman." *Scientific American,* 97, no. 6 (August 10, 1907): 102.

——"An Arctic Day and Night." *McClure's Magazine* 14, no. 6 (April 1900): 555–63.

——"By Airship to the North Pole." *McClure's Magazine* 29, no. 2 (June 1907): 189–200.

——"The Home-Coming of Roosevelt." *American Review of Reviews* 41, no. 5 (May 1910): 555–60.

——"Long-Distance Balloon Racing." *McClure's Magazine* 17, no. 3 (July 1901): 203–14.

——"The Mystery of the Great White North." *Collier's* 30, no. 2 (October 11, 1902): 10–11.

——"On the Way to the North Pole: The Wellman Polar Expedition." *The Century* 57, no. 4 (February 1899): 531–37.

——"The Polar Airship." *National Geographic Magazine* 17 (April 1906): 208–28.

——"Race for the North Pole." *McClure's Magazine* 14 no. 4 (February 1900): 318–28.

——"Sledging toward the Pole." *McClure's Magazine* 14, no. 5 (March 1900): 405–14.

——"The Thousand Mile Voyage of the Airship *America*." *Hampton's Magazine* 25, no. 12 (December 1910): 733–47.

——"The Thousand Mile Voyage of the Airship *America*." *Hampton's Magazine* 26, no. 1 (January 1911): 3–16.

——"The Wellman Polar Expedition." *National Geographic Magazine* 10, no. 12 (December 1899): 481–505.

——"Where is Andrée?" *McClure's Magazine* 10, no. 5 (March 1898): 422–26.

——"Will the 'America' Fly to the Pole?" *McClure's Magazine* 29, no. 3 (July 1907): 229–45.

——"The Wellman Polar Expedition." *National Geographic Magazine* 10, no. 9 (September 1899): 361–62.

Pamphlets

Reymert, Per Kyrre, *Ny-Ålesund: The World's Northernmost Mining Town* (Longyearbyen, Norway: Sysselmannen PA Svalbard, Norway, N-9171, 2016).

Bjerck, Hein B. and Leif Johnny Johannssen, *Virgohamna: In the Air Toward the North Pole* (Longyearbyen, Norway: Sysselmannen, Governor of Svalbard—Environmental Section. N-9171, 1999).

Film and Television

Amundsen: The Greatest Expedition. Directed by Espen Sandberg (Samuel Goldwyn Films, 2021).

The Red Tent (about the *Italia* disaster, starring Sean Connery). Directed by Mikhail Kalatozov (1969).

Roald Amundsen—Ellsworth's Flyveekspedition 1925. (DVD Documentary). Directed by Paul Berge and Oskar Omdal. Norway, 1925.

The Flight of the Airship Norge Over the Arctic Ocean, 1926. (Luftskibit Norge's flugt over Polhavet). Norwegian with English subtitles. Photography and cinematography by Paul Berge and Emil Andreas Horgen. Nasjonalbiblioteket/The National Library of Norway, 1926.

WEBSITES

Airships.net: https://www.airships.net
Embry-Riddle Aeronautical University Intro to Aerospace Flight Vehicles: https://eaglepubs.erau.edu
Flying Whales: https://www.flying-whales.com/en/home/
Hybrid Air Vehicles (HAV): https://www.hybridairvehicles.com
Italia: The Airship Crash Chronicle: https://italia.tass.com
Lighter Than Air Research (LTA): https://www.ltaresearch.com
Roald Amundsen House and Museum: https://amundsen.mia.no/en/

POLAR AERIAL EXPEDITIONS AND CREWS

WELLMAN *CHICAGO RECORD-HERALD* POLAR EXPEDITION (1907)

CREW OF THE *AMERICA*
Walter Wellman, expedition leader
Felix Riesenberg, navigator
Melvin Vaniman, engineer

WELLMAN *CHICAGO RECORD-HERALD* POLAR EXPEDITION (1909)

CREW OF THE *AMERICA*
Walter Wellman, expedition leader, navigator
Melvin Vaniman, engineer
Louis Loud, engineer Nicholas Popov, principle financier, assistant

THE N24/N25 FLIGHT TOWARD THE NORTH POLE (1925)

CREW OF THE DORNIER-WALS FLYING BOATS
N25:
Roald Amundsen, navigator
Hjalmar Riiser-Larsen, pilot
Karl Feucht, mechanic
N24:
Lincoln Ellsworth, navigator
Leif Dietrichson, pilot
Oskar Omdal, mechanic

The Amundsen-Ellsworth-Nobile Transpolar Flight (1926)

Crew of the *Norge*

Roald Amundsen, expedition Leader

Lincoln Ellsworth, expedition co-leader and principle financier

Umberto Nobile, commander of the airship

Hjalmar Riiser-Larsen, first navigating officer

Emil Horgen, second navigating officer

Birger Gottwaldt, wireless operator

Oscar Wisting, helmsman at elevator wheel

Finn Malmgren, meteorologist

Natale Cecioni, chief engine mechanic

Ettore Arduino, assistant chief engine mechanic

Attilio Caratti, engine mechanic

Vincenzo Pomella, engine mechanic

Oskar Omdal, engine mechanic

Renato Alessandrini, rigger and helmsman

Frithjof Storm-Johnsen, wireless operator

Fredrik Ramm, journalist

The *Italia* Expedition (1928)

Crew of the *Italia*

Umberto Nobile, captain and commander of the airship

Adalberto Mariano, first officer

Filippo Zappi, navigator and watch officer

Alfredo Viglieri, lieutenant commander

Felice Trojani, engineer/mechanic

Natale Cecioni, chief technician

Ettore Arduino, chief engineer

Attilio Caratti, mechanic

Vincenzo Pomella, mechanic

Renato Alessandrini, foreman rigger

Calisto Ciocca, mechanic

Giuseppe Biagi, wireless operator

Ettore Pedretti, wireless operator

Ugo Lago, journalist

Dr. Aldo Pontremoli, scientist

Dr. Francis Behounek, scientist

Dr. Finn Malmgren, meteorologist

GLOSSARY OF AIRSHIP AND AVIATION TERMINOLOGY (1900–1930s)[1]

AFT: Rear of an airship.

AIRSHIP: The general term for a powered, steerable lighter-than-air vehicle, such as a balloon, blimp, or zeppelin. "Airship" is synonymous with "dirigible." There are three primary types, defined by their method of maintaining their shape: non-rigid, semi-rigid, and rigid. Non-rigid and semi-rigid airships are referred to as "pressure airships." Walter Wellman's original *America* (1906) was a non-rigid airship; his subsequent versions were stiffened by the addition of an attached keel. The *Norge* and *Italia* were semi-rigid. A general rule (though there are exceptions) is that non-rigids are small, semi-rigids are medium-sized, and rigids are large. Zeppelins were rigid airships.

AIRSPEED: Speed relative to the air, independent of the effects of wind.

AMIDSHIPS: Halfway (midway) between the bow and stern of an airship (or sailing/steam ship), either longitudinally or laterally.

---

1 The terms found in this glossary were gleaned, confirmed, and clarified from a variety of sources, including the following: Embry-Riddle Aeronautical University Intro to Aerospace Flight Vehicles: https://eaglepubs.erau.edu; The Ligher-Than-Air Society: https://www.blimpinfo.com/learn/fun-facts/glossary/; Airships.net https://www.airships.net; Airship Saga: The History of Airships through the eyes of the Men Who Designed, Built, and Flew Them by Lord Ventry (Blandford Press/Sterling Publishing. London: 1982); Giants in the Sky: A History of the Rigid Airship by D. H. Robinson (Seattle: University of Washington Press, 1973); and two excellent books by Garth James Cameron: *From Pole to Pole: Roald Amundsen's Journey in Flight* (Skyhorse Publishing. New York: 2013) and *Umberto Nobile and the Arctic Search for the Airship* Italia (Fonthill Media Limited, 2017).

ANGLE OF ATTACK: The angle between the fore and aft axis of an airship and the relative air flow. An airship flying nose up has a positive angle of attack; an airship flying nose down has a negative angle of attack.

ARTIFICIAL HORIZON: Instrument used to indicate an airship's (or any aircraft's) orientation relative to the earth, expressed as yaw, roll, and pitch. Assists in low visibility or bad weather conditions when the actual horizon is not visible as a reference for the pilot.

AUTOMATIC VALVE: Valves at the top of an airship, often spring-loaded, which automatically open to release gas when the pressure in a gas cell exceeds a predetermined level.

BALLAST: Weight, typically in the form of water or sand (but also metal shot pellets or balls carried in canvas bags), used to offset the buoyancy of an airship's lifting gas (hydrogen or helium). Ballast could be dropped or "jettisoned" if the airship became too heavy, or to aid in landing.

BALLAST CHAIN: A heavy cable or chain, attached by one end to an airship, lowered to ease the descent of an airship.

BALLOON: An unpowered, lighter-than-air craft or vehicle. Balloons get their buoyancy from confined hot air, hydrogen, helium, or another gas. Traditionally spherical in shape, like an inverted teardrop.

BALLONET: An independent air bag (bladder) within the envelope of a pressure airship (non-rigid or semi-rigid). A ballonet is inflated or deflated to preserve a continuous volume in the envelope as the lifting gas volume increases or decreases. A ballonet—or more often numerous ballonets—maintain the shape of the airship, ensuring control and performance.

BAROGRAPH: An onboard altimeter and chronometer that recorded altitude and time.

BLIMP: Small pressure airships were called "blimps" as early as 1915. The term is believed to have been coined in describing the sound produced when one snapped their thumb or finger on an airship's fully inflated envelope.

BUMPER: An air-inflated, shock-absorbing rubber and/or canvas structure located beneath the control car to soften landings. Umberto Nobile used them on the *Norge* and *Italia.*

CELESTIAL NAVIGATION: Determining one's position by observing celestial bodies such as the sun, moon, or stars.

CONTROL CAR: Also referred to as a gondola or control gondola. Enclosed structure attached beneath the underside of an airship's envelope (typically forward) that houses the captain, navigator(s), radio operator(s), and all airship controls and instruments. Passengers and observers also rode in the control car.

CONTROL SURFACES: Surfaces or structures such as the rudders (attached to the vertical fins) to control the flight of an airship. Elevators attached to horizontal fins controlled pitch.

DEAD RECKONING: Using wind speed and direction to calculate a current position relative to a previously determined position or "fix."

DIRIGIBLE: General term for airship but often used to refer to larger airships (semi-rigid and rigid).

DRIFT: Sideways motion of an airship relative to a prevailing wind.

DYNAMIC LIFT: Lift or vertical movement of an airship resulting from air flowing over the vehicle (aerodynamic forces acting on the shape of the airship), as opposed to "static lift," which is lift generated by the buoyancy of the lighter-than-air lifting gas (typically hydrogen or helium).

ELEVATORS: Surfaces that control the airship's pitch (nose up or down), usually attached to the horizontal fins.

ENGINE CAR: Compartment that contains an engine (or engine and mechanic/operator). Sometimes also referred to as an engine boat or engine gondola or engine pod.

ENVELOPE: The bag containing the lifting gas (hydrogen or helium) in a pressure non-rigid or semi-rigid airship. Also called a gas bag. Originally fabricated of layers of rubberized cotton fabric, which were "doped" with plasticized lacquer for durability and protection against the elements and abrasion.

EQUILIBRIUM: The state of an airship when it neither rises nor descends, and when the forces of lift and gravity are equal.

FINS: Horizontal and vertical surfaces attached to the rear of an airship for stabilization of pitch (nose up and down) and yaw (left and right) movement.

FIX: A calculated position that is reliable. Fixes could be obtained by observation of surface features or landmarks (such as mountains, rivers, lakes), by successive sun sights, or from the intersection of position lines obtained by direction finding two or more radio stations simultaneously.

FORE: Front of an airship.

FREE BALLOONING: When an airship's engines are stopped, and it drifts with the wind.

FLYING BOAT: A seaplane (airplane) employing a waterproof, boat-shaped fuselage for buoyancy on water. The Dornier-Wals *N24* and *N25* used by Roald Amundsen were flying boats.

GAS VALVE: Valves used to vent lifting gas. These could be automatic (emergency) or manual (controlled by captain or pilot).

HANGAR: Shed constructed to house or store an airship. Also called a shed or hall.

HEAVIER-THAN-AIR AIRCRAFT: Airplanes and gliders, which garner their lift by air moving over the wings in forward flight.

HYDROGEN GAS: The lightest of all gases and relatively inexpensive to produce (especially compared to helium), hydrogen gas was used to inflate balloons and airships. It is explosive when mixed with oxygen. The *America, Norge,* and *Italia* were all inflated with hydrogen gas.

KEEL: A stiffening structure running beneath the underside of a semi-rigid airship. In both the *Norge* and *Italia,* the control car and engine cars were affixed to the keel and could store fuel and equipment. A walkway or "catwalk" ran along the bottom of the keel the entire length from tail to nose.

LIFTING GAS: Hydrogen or helium. Wellman's *America* and Nobile's *Norge* and *Italia* were all inflated with hydrogen as their lifting gas.

LIGHTER-THAN-AIR AIRCRAFT: Aircraft such as balloons and airships that are lifted by the buoyancy produced by the lifting gas within their envelopes.

MOORING MAST: A tower to which an airship could dock, particularly in high winds.

NACELLE: Another term for engine car, engine gondola, or engine boat. External housing for an engine and engine mechanics.

NOSE CONE: Reinforcing structures of metal tubing to strengthen the nose sections of airships, which is especially useful when docking to mooring masts.

PITCH: The movement of an airship's nose up or down, controlled by the elevators.

PORT: Left.

PRESSURE AIRSHIP: An airship whose shape is maintained by keeping the gas within the envelope at a higher pressure than that of the surrounding atmosphere. Non-rigid and semi-rigid airships are pressure airships.

PRESSURE HEIGHT: The altitude at which a pressure airship's envelope (or the gas cells of a rigid airship) are 100 percent full. If the airship rises beyond pressure height, gas must be valved off to avoid the rupturing of envelope or gas cells.

RIGID AIRSHIP: An airship constructed with a rigid framework covered by fabric, containing a number of gas bags or ballonets inside. The framework (rather than the ballonets) maintains the airship's shape. The *Hindenburg, Graf Zeppelin,* and USS *Shenandoah* were all rigid airships.

RIP PANEL: A panel in the envelope of a pressure airship that will open when a rip cord is pulled, letting out gas immediately. Can be used in an emergency or when landing in strong winds.

ROLL: Movement of an airship about its fore and aft axis.

RUDDER: Control surface connected to an airship's vertical fin allowing control of yaw (to move the nose left or right).

SEAPLANE: Airplanes such as float planes and flying boats designed to take off from and land on water.

SEMI-RIGID AIRSHIP: Pressure airships using a keel along the envelope's bottom to help stiffen and maintain the airship's shape. Control cars or gondolas could be attached below the keel. The *America, Norge,* and *Italia* were all semi-rigid airships.

STARBOARD: Right.

STATIC LIFT: Lift created by the airship's lifting gas (hydrogen or helium).

SUN COMPASS: A navigational tool that uses the shadow of the sun to sustain direction. Useful when magnetic compasses become unreliable, as when flying near the magnetic poles, which create anomalies and disturbances.

TOTAL LIFT: Dynamic and static lift combined equal total lift.

TRIM: The fore and aft position of an airship. An airship is "in trim" when level and "out of trim" when the nose is either up or down. Airship trim can be adjusted by moving cargo weight fore or aft, or by altering the amount of air in fore and aft ballonets.

USEFUL LIFT: The amount of lift available for fuel, cargo, ballast, passengers, and crew.

VARIOMETER: Instrument to determine the rate at which an airship is ascending or descending (typically expressed in feet per minute).

YAW: The movement of an airship's nose left or right.

ZEPPELIN: Often a default term to describe any large airship, zeppelins were rigid airships designed by the Zeppelin Company in Germany.

INDEX

ABOUT THE AUTHOR

Alaina Mullin

BUDDY LEVY is the author of more than half a dozen books, including *Empire of Ice and Stone: The Disastrous and Heroic Voyage of the* Karluk; *Labyrinth of Ice: The Triumphant and Tragic Greely Polar Expedition; Conquistador: Hernán Cortés, King Montezuma, and the Last Stand of the Aztecs;* and *River of Darkness: Francisco Orellana and the Deadly First Voyage Through the Amazon.* He is coauthor of *No Barriers: A Blind Man's Journey to Kayak the Grand Canyon* and *Geronimo: Leadership Strategies of an American Warrior.* His books have been published in a dozen languages and won numerous awards. He lives in Idaho.